PLANNING FOR A NEW ENERGY AND CLIMATE FUTURE

Scott Shuford, AICP, Suzanne Rynne, AICP, and Jan Mueller

TABLE OF CONTENTS

Chapter 1: Introduction 1

Chapter 2: Energy 5

Chapter 3: Climate Change 19

Chapter 4: Greenhouse Gases 39

Chapter 5: Strategic Points of Intervention 47

Chapter 6: Development Patterns 61

Chapter 7: Infrastructure and Utilities 73

Chapter 8: Transportation 87

Chapter 9: Economic Development 101

Chapter 10: Buildings and Site Design 119

Chapter 11: Natural Resources 133

Appendix: Two Energy Surveys 149

References 151

CHAPTER 1

Introduction

 Today's global climate change and energy challenges include the need to promote greater energy efficiency and use of renewable energy, reduce greenhouse gas emissions, and prepare for and adapt to a changing climate. To address these complex issues, community planners must understand the scientific basis for taking action and utilize the most effective planning tools and techniques for the specific challenges they face. Taking action could help reduce our reliance on nonrenewable energy sources, help communities better meet their energy needs, improve environmental quality, and generate other benefits such as improved health, quality of life, and increased investment in the local economy. In short, there are numerous benefits to taking many of the actions recommended in this report, even if energy or climate concerns are not motivating factors in a given community.

> **ASSESSING THE INTEGRATION OF ENERGY AND CLIMATE INTO PLANNING**
>
> In 2005 and 2007, APA conducted surveys to assess the awareness of planners about energy and climate-change issues and the integration of these issues in planning. The results indicated an increasing awareness that energy and climate issues were important. However, the findings also indicated the need for more and better tools, techniques, and strategies for integrating energy and climate into planning. A summary of the findings from these surveys is in the appendix.

Many individual American cities, towns, and counties—whether urban, suburban, or rural—are already doing a great deal of positive work on addressing climate change and on promoting energy efficiency and the use of renewable energy sources. To a great extent, the battles to adapt to and mitigate the effects of climate change and to reduce our dependence on greenhouse gas–producing energy sources will be won or lost not at the state or national levels but closer to home. While states are producing energy plans and climate action plans, and the federal government is working toward climate and energy solutions, there are broad opportunities for effective action at the local and regional level. Moreover, work at these levels will serve to implement and complement state and federal policies.

Why Planners?

Planners have an important role to play with regard to energy and climate change challenges. Their education and experience affords them a skill set particularly useful for analyzing complex problems and opportunities and for developing strategies that address them.

- The comprehensive perspective that planners have is particularly helpful in understanding how energy and climate relate to and affect other issues, such as land use, economic development, and transportation. Planners often work with and understand aspects of many different disciplines, allowing them to identify opportunities for synergy and interrelationships in plans and implementation.

- Planners have a long-range outlook. They are trained to look at changing conditions and plan 5, 10, 20, or more years into the future, which is important when thinking about potential long-term impacts from climate change and in making investments in energy systems and processes.

- Planning is one of the few professions that focuses on place-based problems and opportunities affecting health, safety, and general welfare. Planners have a particular ability to deal with the community-wide spatial component of environmental, infrastructure, public safety, and quality-of-life issues.

- Planners are trained to spot and deal with unintended consequences and long-term cumulative impacts. This is particularly important in adapting to possible impacts from climate change and identifying sustainable energy solutions.

- Planners have special expertise in community engagement and consensus building. They often act as conveners of stakeholders. They can play an important role in involving a community in discussions about taking actions to address and respond to climate change.

- Planners are often strategically well placed within a city, town, or county to take a collaborative or leading role on such issues. This work often requires cooperation and support from a wide variety of agencies and departments, such as public works, engineering, parks and recreation, the transit authority, the housing authority, and utilities, as well as local NGOs. The planning director often has working relationships with all these entities and knows their languages.

Because of planners' strategic centrality on these topics, mayors, city managers, and other local policy makers often turn to them for help in addressing these complex problems. This makes it extremely important that public and private planners at all career stages have the best scientific information available to make the case for engaging in sound climate and energy planning, as well as guidance on how to integrate climate change and energy issues into all stages and phases of planning.

Planners have long advocated for community actions that resonate well with actions necessary to respond to climate and energy challenges. Efforts to combat sprawl through promotion of a more compact land-use pattern and mixed uses, for instance, can result in fewer vehicle-miles traveled, reducing the amount of greenhouse gases going into the atmosphere. Planners have advocated for broadening transportation choices through the development of bus and light-rail transit, bicycle and pedestrian facilities, and complete streets policies. But there have been few communities that have made a concerted and comprehensive effort to integrate climate change and energy issues into all stages of planning and community development. Now that climate change and energy have become such high-priority global issues, planners must make climate and energy yet another driving force for good planning.

Organization

This report is organized in two sections. The first section, Chapters 2–4, presents fundamental information about energy and climate change. Readers will find information on energy efficiency and renewable energy, an overview of the science of climate change, and its anticipated impacts. A discussion of greenhouse gas emissions and how to measure a local greenhouse gas footprint follows.

The second section, Chapters 5–11, focuses on what planners can do to plan for energy and climate change. Chapter 5 outlines the strategic points of intervention—that is, places in the planning process where planners can make a difference—while Chapters 6 through 11 describe ways to consider energy and climate concerns in six different issue areas that planners often address. Tables at the end of each of these chapters provide examples of actions within each issue area that correspond to the strategic points of intervention. Case studies are used throughout to showcase communities that have already begun to take some of the steps discussed here.

More case studies and resources are available in an online, searchable database at www.planning.org/research/energy/database/index.htm, which is a complementary resource to this report.

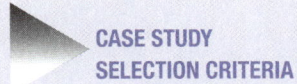

CASE STUDY SELECTION CRITERIA

In selecting case-study communities, researchers considered the following criteria:

Balanced Coverage
- *Geography*, in order to highlight different U.S. regions
- *Community Size*, using small communities, large cities, and those in-between
- *Scale*, using local, regional, and even state examples

Excellence in Practice
- *How strongly it relates to planning practice and the points of intervention*—is it an example that planners would find useful?
- *Issues it covers*—does it address issue areas that planners typically confront?
- *Implementation*—what has been implemented or what plans are there for implementation?
- *Community and stakeholder involvement*—who else was involved?

CHAPTER 2

Energy

 Energy is the lifeblood of modern societies. The economic expansion of the 20th century in the United States and other industrialized countries was powered by abundant, inexpensive energy, primarily from fossil fuels: coal, oil, and natural gas. The questions for planners and policy makers are: how will the 21st century be different, how will our energy needs be met, and how much will energy cost?

GLOBAL UNCERTAINTIES AND LOCAL STRATEGIES

Energy forecasts by even the most authoritative agencies and organizations rely by necessity on tenuous assumptions and incomplete information (EIA 2009b). The future of energy, moreover, will ultimately be shaped by human behavior and collective decisions—further adding to the uncertainty. But the world's energy future will almost certainly be different than its energy past (EIA 2009b).

New energy technologies are emerging that are likely to alter where we get and how we use energy, as well as the design and scale of energy systems. Energy development in the 20th century was dominated by fossil fuels and centralized energy sources: large power plants, large petroleum facilities, large hydroelectric dams, and so on. New energy technologies and the desire to provide more secure and sustainable energy systems may drive new business models in the 21st century. The prospects of greater energy efficiency, renewable energy sources, and decentralized energy systems offer local communities opportunities to prepare for change and to shape their own energy futures. Planners are well situated to lead the development of local and regional energy strategies. A local strategy does not replace but rather complements state and national policies; indeed, local action may be essential to the success of broader energy policies and plans. Local strategies are informed by an understanding of local energy resources as well as the context of larger national and international energy issues. This chapter provides a brief overview of current energy trends.

GLOBAL SUPPLY AND DEMAND

Global energy use has grown steadily over the past century as both world population and economic activity has expanded. Energy production has kept pace with expanding demand during this time without significant increases in real prices; in some cases, real prices have decreased (EIA 2008). Various studies and analyses have examined whether that trend will continue or whether production capacity of one or more fossil fuels, particularly oil, may reach a peak or level off (Wood et al. 2004, Energy Watch Group 2007). The world is not expected to run out of oil, coal, or natural gas anytime soon, but unprecedented global demand and gradual depletion of the most accessible and least costly fossil-fuel reserves may significantly affect both production costs and basic economics of supply and demand—resulting in higher prices, increased price volatility, or both. Experts disagree on when or how this scenario is likely to unfold, but most concur it is an inevitable result of reliance on finite, nonrenewable energy resources.

Developed and industrialized countries accounted for a majority of the rising energy demand over the past century. Energy demand in newly industrializing countries—such as China, India, and Brazil—are expected to drive and accelerate growth in energy demand in the coming decades (EIA 2009b).

U.S. ENERGY OUTLOOK

The United States accounts for roughly 20 percent of all global energy consumption, which is presently more than any other single country (EIA 2009b). The U.S. Department of Energy projects U.S. energy demand to grow by 11 percent by 2030, or approximately 0.5 percent per year, based on an extrapolation of current trends (EIA 2009a). Energy use per capita in the United States has actually remained fairly stable over the past few decades, while the energy intensity of the U.S. economy (measured by energy use per dollar of economic output) has steadily declined—about 1 to 2 percent per year—over the last 30 years (EIA 2009a). (See Figure 2.1.) Population increases and economic growth, however, have outstripped those efficiency gains. A significant but unknown variable in these "business-as-usual" projections is the extent to which future energy needs will be met by greater energy efficiency or by new energy resources.

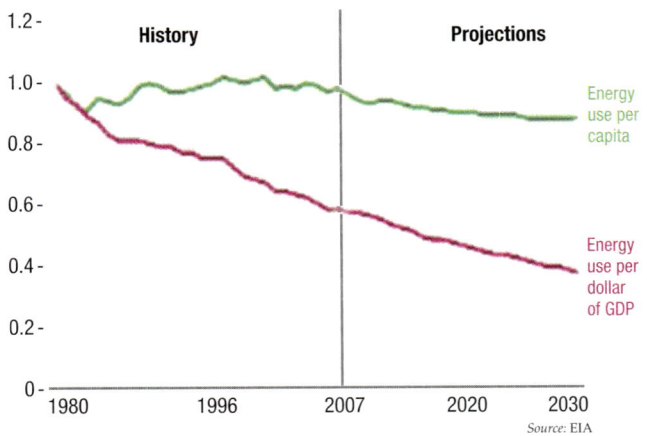

Figure 2.1. Energy use per capita and per dollar of GDP, 1980–2030 (1980 = 1)

EXISTING ENERGY SOURCES

Fossil fuels currently dominate U.S. energy sources. Of the 102 quadrillion BTUs or "quads" of total primary energy consumed in the United States in 2007, 85 percent came from oil, natural gas, or coal. Oil and fuels derived from petroleum are, by far, the largest source—providing 40 percent of all energy consumed. Natural gas and coal compete for second place at 23 and 22 percent of total primary consumption. Nuclear power plants (8 percent) and renewable energy sources (6 percent) make up the remainder. Among renewable sources, discussed below, all uses of biomass combine to provide 3 percent of total primary energy, and biomass now exceeds hydropower (2 percent) as the largest renewable source. Other renewable options including wind, solar, and geothermal sources currently account for about 1 percent of all U.S. energy consumption (EIA 2009a).

While it is helpful to compare overall sources of consumed energy, it is important to note that different energy sources are used disproportionately for different purposes. Oil and petroleum-based fuels are used largely for transportation and industrial uses. Coal is used primarily to generate electricity. Natural gas is more versatile and is used for electric generation, many industrial purposes, and space and water heating. Nuclear energy, hydropower, solar energy, and wind energy are used primarily to make electricity. Biomass is currently being used or processed for heating, electricity, and transportation fuel purposes. The future may see more interchangeability and shifts among sources for different uses—increased use of electricity for transportation, for example—but it is helpful to keep in mind the opportunities and limitations of different energy sources for different end uses.

Approximately 40 percent of primary energy is used to generate electricity. Other sources are generally either combusted directly or consumed in industrial processes. A brief overview of major energy end uses provides a basic sense of where all the energy goes. All figures regarding end uses are derived from the Annual Energy Outlook 2009, published by the U.S. Energy Information Administration (EIA 2009a).

ENERGY USES

Industry

The U.S. industrial sector—including electricity used for industrial purposes—is the largest energy user, relative to the other three major sectors: transportation, residential, and commercial uses. Industry accounts for one-third of all U.S. energy use, including approximately half of all natural-gas use, one-third of all oil consumption, and 27 percent of all electricity use (EIA 2009a).

Manufacturing of all types is the dominant category within this sector, accounting for 85 percent of the total. The U.S. Department of Energy includes nonmanufacturing uses such as agriculture, mining, and construction under its definition of industrial uses, but these constitute a minor portion. Within manufacturing, chemical production, refining, and iron and steel production are the largest users, but the amount of energy used for these purposes is still exceeded by all other manufacturing activities.

Transportation

Transportation is the second-largest energy consumer among the four major sectors, accounting for nearly 30 percent of all primary energy use and more than 70 percent of all oil consumption. Transportation also uses about 4 percent of all direct natural-gas consumption and approximately 2 percent of all electricity.

Light-duty cars and trucks account for about 60 percent of all transportation-related energy use, with freight trucks accounting for 20 percent and air travel accounting for another 10 percent. The remainder is shared among shipping (4 percent), rail (2 percent), and other uses.

Residential and Commercial Sectors

The residential and commercial sectors use less energy overall than the other two sectors (approximately 22 percent and 19 percent, respectively), but these sectors together account for the majority of electricity use and account for one-third of all natural-gas consumption.

These two sectors, while different in many ways, share a number of major energy uses—chiefly space heating and cooling, water heating, lighting, and a wide variety of electrical appliances and equipment. These major uses will be examined together to help illuminate opportunities to address them jointly.

Space heating and water heating account for more than 90 percent of the natural gas used in these two sectors. While the use of electricity for heat has declined greatly in the United States over the past several decades, space and water heating still account for more than 10 percent of residential and commercial electricity use. Space cooling, on the other hand, is generally an electric endeavor, consuming about 15 percent of all residential and commercial electricity and more than 10 percent of all electricity consumed in the United States. A minor amount of oil consumption (approximately 2 percent) is associated with space heating, though oil is a more significant heating source in some regions of the country.

Both residential and commercial sectors use a large amount of electricity for different kinds of household appliances and equipment, and this class of electricity use has been among the fastest growing. Household appliances account for roughly half of all residential electricity use, and commercial appliances and equipment account for half of all commercial electricity use. (This excludes associated water heating and lighting uses.) Refrigerators and freezers, computers, televisions, and clothes dryers stand out within this category—accounting for 9, 5, 4, and 3 percent of electricity used by these sectors, respectively—but the total is widely distributed among a plethora of electric and electronic devices. Ventilation of commercial workspaces is also a significant energy user (about 5 percent of electricity used in these two sectors).

Lighting accounts for roughly 20 percent of all residential and commercial electricity use. This includes all indoor and outdoor lighting used for public and institutional purposes. Miscellaneous uses including commercial and residential cooking account for the remainder of natural-gas and electricity use.

NEW ENERGY SOURCES

What are the options to supplement or displace fossil-fuel energy sources, and which ones are feasible for a given community? To help answer those

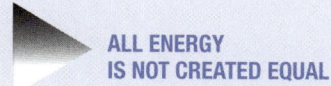

ALL ENERGY IS NOT CREATED EQUAL

In areas most likely to be influenced by local planners, we use energy for three main purposes: transportation, heating and cooling of buildings, and electricity to power everything else (lighting, appliances, etc.). There is considerable overlap among the energy sources used in these applications; natural gas, for example, is widely used for peak electric power, for building heat, and as a fuel for buses and other heavy-duty vehicles. However, all energy sources are not equally favorable for different uses. Electricity, for example, is relatively efficient for producing mechanical power, including transportation motors, but it is terribly inefficient for producing heat. Different energy sources have inherent opportunities and constraints that affect how efficient and cost-effective they are for different uses.

questions, existing and emerging technologies for energy efficiency and renewable energy are reviewed next.

Energy efficiency and conservation measures are often the most cost-effective option for "producing" more energy and, in a vast number of examples, have achieved net cost savings—that is, they paid for themselves and earned money over time.

Studies by the U.S. Department of Energy and other analyses have estimated that the United States could reduce its total energy use by approximately 25 to 30 percent below business-as-usual projections by 2020 to 2030 (NRC 2009; McKinsey and Company 2009) while achieving tens of billions of dollars in cost savings. Other studies have suggested that comprehensive efficiency measures across all sectors of the economy could cut U.S. energy use in half (Lovins 2005). Several other industrial economies (e.g., Japan, Germany) currently use approximately 40 to 50 percent less energy—per capita or per dollar of economic production—than the United States (EIA 2009b).

Energy efficiency can be aided by specific technologies but can also be enhanced through more efficient planning and design that results in more efficient system operations and more energy-saving decisions by businesses and consumers. For example, at the individual level, a person switching from a car with a conventional engine to a comparable vehicle with a hybrid electric engine may reduce his or her fuel consumption by about 15 to 25 percent (EPA 2009a). Alternatively, that same person could move within walking or biking distance of work, join a carpool, or take transit for commuting trips and also reduce fuel consumption by a comparable amount. Commuting trips make up approximately 15 to 20 percent of passenger miles traveled, on average (TRB 2006).

At the community level, the most effective efficiency strategies combine the benefits of technology, planning and design, and incentives for more efficient decisions and behavior by energy users. Planners bring an important perspective in addressing and integrating these different aspects of energy efficiency. But there is no silver bullet; promoting energy efficiency is a game of finding numerous small measures that add up to big savings. Each community needs to assess its own opportunities, priorities, and information to determine the relative cost-effectiveness and scale of potential gains that can be realized through efficiency. There are, however, several major areas of opportunity to substantially improve energy efficiency that are relevant to most communities. These opportunities will be discussed further within specific issue areas in Chapters 6–11.

At the same time that communities pursue energy efficiency, they can also plan for a transition to new energy sources. Renewable energy technologies, in particular, present opportunities for local communities to develop local and regional energy resources and address their own local energy needs.

Electricity

Biomass and waste-to-energy. Biomass energy production involves the use of organic matter (also known as biomass) to produce electricity, heat, or transportation fuels (biofuels). Biomass sources, commonly referred to as "feedstocks," include a wide variety of materials such as agricultural residues, low-grade timber, woody debris, waste wood, food waste, farm animal manure, and energy crops. Burlington, Vermont, for example, gets approximately two-thirds of its electricity from a power plant fueled almost entirely with wood chips. The Netherlands has constructed a combusting facility that converts 25 percent of the chicken manure generated in the country to electric energy for 90,000 homes.

The cost of electricity from biomass can vary with location and type of feedstock. In some cases the choice of feedstock may reduce waste-disposal costs.

Figure 2.2. A gas compressor island at a biomass gasification facility in Maui, Hawaii. The biomass gasifier uses residue from the nearby sugarcane mill.

As in any energy production, the efficiency of the equipment and the amount of emissions it generates is an essential consideration; major advances have been made in both areas in recent years. The Oregon Department of Energy reports biomass-generation costs of approximately five to seven cents per kilowatt-hour, compared to about three cents per kilowatt-hour for a new advanced gas-fired power plant (Oregon Department of Energy n.d.). Biomass also can be burned in conjunction with other fuels, such as coal, to lower emissions.

Municipal solid waste (MSW) is composed of a mixture of organic materials, as well as metals, plastics, and other petroleum-derived products. Waste-to-energy applications can lower GHG emissions (relative to fossil fuels) and reduce landfill loading (along with associated methane emissions). For example, Warrenton, Virginia, is planning to build a gasification plant that will use the town's municipal waste to meet almost all of its electric needs—thus reducing both emissions and the volume of unreclaimed waste.

Actual GHG reduction benefits of different bioenergy and biofuels technologies can vary significantly depending on which feedstocks are used and what methods are used to grow, collect, and process the fuel. The indirect land-use effects of diverting food crops for energy production—whereby new acres are cultivated as a result—may also offset the GHG benefits of some biofuels. Some studies (Searchinger et al. 2008) have suggested these effects may entirely negate GHG benefits and increase overall GHG emissions. Nevertheless, there are many biofuels and bioenergy feedstocks that, by contrast, can be harvested from waste streams and by-products of agricultural and forestry operations.

Wind. Large-scale wind-power installations or "wind farms" using large turbines that can each generate more than five megawatts of electricity have been built or are being explored in many states. "Small wind" installations, using individually owned and smaller turbines that generate fewer than 100 kilowatts of energy, can also be part of a local energy strategy. Several communities around the country have also pursued "community" wind options that are intermediate in scale. Community wind projects may serve from a few to a few hundred households and businesses. The American Wind Energy Association website (www.awea.org) contains information about wind turbines in various applications, including at the single-family residential level. Windustry, a nonprofit organization based in Minneapolis, has been a leader in pioneering community approaches to wind energy.

Total costs and energy savings of installing wind turbines can vary depending on the location and scale of the project, but wind energy is becoming competitive with coal- and gas-fired power plants in many places, with costs in the range of five to 10 cents per kilowatt-hour (AWEA n.d.). Nationwide, net electricity generation from wind grew by 50 percent in 2008 (AWEA 2009). Recent studies conservatively estimate that wind energy could meet 20 percent of U.S. electricity demand by 2030 (NREL 2007).

Large-scale wind turbines can have visual impacts due to their height. Some of the largest turbines have rotor diameters of more than 110 meters. Optimal location and height is highly site-dependent. Additional community concerns may include effects on wildlife and noise, although industry studies have indicated that these effects are largely minimal. Small-scale wind turbines also require locations that are typically highly visible in order to have unobstructed access to wind. While new turbine technologies that can capture wind from varying directions are making it easier to generate wind power in urban areas, adjustments to zoning ordinances and accessory-use requirements may be necessary to accommodate small wind turbines.

Figure 2.3. Wind turbines in Buffalo Ridge, Minnesota

Solar. Solar energy encompasses passive solar (such as water and space heating) and active solar, which generates electricity. Solar electric technologies include photovoltaic panels that convert solar energy directly to electricity and concentrated solar power (CSP), which uses mirrors to focus solar rays in order to heat fluids that drive a turbine. The cost-effectiveness of electricity produced by photovoltaic panels is more favorable at lower latitudes, where solar radiation is more intense, but solar energy has proven practical and feasible in northern regions with relatively cloudy weather patterns. Solar energy can take advantage of solar power in a wide variety of local sites, from roofs to parking lots. Costs per kilowatt-hour of delivered electricity can vary widely from project to project depending on location, labor costs, project size, and other factors, but prices for solar panels themselves are trending downward, and average prices are getting closer to being competitive with other conventional sources (Solar Buzz Consultancy 2009).

Figure 2.4. Sacramento Municipal Utility District's photovoltaic array

Passive and active applications can work with both new construction and building retrofits. Solar-access rights for residential solar panels need to be defined in local codes to avoid conflicts among property owners.

Applications for solar power are not limited, and rapidly developing technologies are making solar an increasingly viable alternative-energy option. The old style of silicon-based photovoltaic cells that make electricity from the sun's energy puts the price of solar power at around $3 to $4 per watt; the current price of most newer solar-energy technologies is around $1 per watt. However, several companies are now perfecting thin-film solar technology that is more versatile in its applications, is less chemical- and energy-intensive to produce, and can produce electricity much more cheaply than current technology can. In a recent interview, Anil Sethi of Flisom, a Swiss manufacturing firm, foresees his company's thin-film solar panels reaching $0.80/watt in five years and $0.50/watt within 10 years (Gordon 2007).

In 2008, the Sharp Corporation introduced its new system for focusing sunlight with a Fresnel lens (like those used in lighthouses) onto superefficient solar cells, which are about twice as efficient as conventional silicon cells (Bullis 2006). The eSolar Corporation utilizes arrays of mirrors to concentrate solar energy on water-filled boiler units that create steam for turbine-generated electricity; their standard 46 megawatt array can power 30,000 homes yet requires only about one-quarter of a square mile.

Solar space and water heating. Solar hot-water systems, which are used to heat water for domestic and commercial applications, are one of the most widely used and cost-effective renewable-energy technologies. The technology is relatively simple, does not require especially sunny weather or locations to operate, and can pay for itself in energy cost savings in fewer than 10 years (Renewable Energy Resource Center n.d.).

Businesses such as the Proximity Hotel in Greensboro, North Carolina, are investing in solar hot-water systems to save money and reduce their environmental impact. The 100 panels on the hotel's roof provide 60 percent of the hotel's hot water and saved the hotel approximately $14,000 in one year (Rowe 2006).

Hydropower. Hydropower includes electricity generated by hydroelectric dams, but there are also other ways of harnessing power from moving water such as tides, waves, and free-flowing (undammed) rivers.

According to the U.S. Department of Energy, hydropower plants can be characterized by the scale of their generation of electricity:

- Large hydropower: greater than 30 megawatts
- Small hydropower: 100 kilowatts to 30 megawatts
- Microhydropower: fewer than 100 kilowatts

Large and small hydropower systems typically require the construction of an impoundment or dam to allow the moving water to be managed for electrical generation purposes. Microhydropower may use impounded or diverted water.

Hydroelectric power has many advantages as a sustainable low-carbon alternative to fossil fuel–derived electricity generation. It uses an existing technology to exploit a clean, renewable energy source. Moreover, this energy can be produced on demand by regulating the flow of water through the turbines. This, however, requires impoundment of water in a reservoir, which can result in:

- Flooding of large areas of land upstream of the dam, destroying existing habitat areas.
- Slowing the rate of flow of a river, which can cause water temperatures to rise, sometimes well above natural temperatures.
- Temperature stratification. The depth of the reservoir created by the dam creates cold water, which contains less dissolved oxygen; releasing this colder water downstream affects species that may not be able to survive in a less oxygen-rich environment.
- Disruption of natural sedimentation and flood patterns, thus altering the upstream and downstream ecosystems. Nutrient-rich sediment becomes trapped behind the dam, causing nutrient loading and depletion of dissolved oxygen. Downstream, less sediment is deposited; without depositions of fresh sediment, riparian plants are unable to grow on banks, causing erosion.
- Interruption of fish migration patterns.
- Tectonic and groundwater impacts, notably with large impoundments. (Foundation for Water and Energy Education n.d.)

Further, in drought-prone areas, water levels may fall, decreasing the amount of electricity that can be generated. Many hydroelectric plants in the southeastern United States, for example, have been recently in danger of having to suspend operations due to low water levels. During a 2007 drought, the Army Corps of Engineers Savannah District showed that hydroelectric power production was 5,000 to 6,000 megawatt-hours per week below average (Kitzmiller 2007). Drought conditions may be worsened by climate change, reducing hydropower's benefit as a reliable source of energy.

On the other hand, studies show the potential to draw 5,000 megawatts from new hydropower installations at existing nonpowered dams. Additionally, greater efficiencies can increase the generating capacity of many existing hydropower facilities (Electric Power Research Institute 2007).

Microhydropower systems. Structures that have access to moving water may be able to utilize small-scale hydropower systems. Such systems are capable of meeting electric power needs of a building or even a complex of buildings. In suitable locations, microhydropower can provide an inexpensive, reliable source of electricity for distributed generation (Appalachian State University 2007) and may not require the construction of a dam or impoundment.

More information on hydropower can be found on the U.S. Department of Energy's Energy Efficiency and Renewable Energy website: www1.eere.energy.gov/windandhydro.

Tidal and wave energy. This technology harnesses the kinetic energy in large bodies of water and transforms it into electrical energy. According to the World Energy Council, wave power has the potential to supply 10 percent of the world's energy (at current consumption levels). Renewable-energy analysts at the U.S. Department of Energy estimate that current technologies could be used to recover up to two terawatts of electricity from the world's oceans (DOE 2008).

Geothermal. Electricity can be produced using hot water found in deep geologic formations. This is generally an industrial-scale operation and is limited to select areas. A 2008 U.S. Geological Survey study found that the United States has significant power-generation potential from the development of geothermal systems (USGS 2008).

A more common and small-scale option, however, is the use of geothermal heat-exchange systems, also known as ground-source heat pumps, which use the relatively stable temperature of the earth to reduce energy use for heating and cooling. Almost everywhere, the upper 10 feet of the earth's surface maintains a nearly constant temperature between 50 and 60°F (10 and 16°C). A geothermal heat-pump system consists of pipes buried in the shallow ground near a building, a heat exchanger, and ductwork into the building. In winter, heat from the relatively warmer ground goes through the heat exchanger into the house. In summer, hot air from the house is pulled through the heat exchanger into the relatively cooler ground. Heat removed during the summer can be used as no-cost energy to heat water.

Transportation

Fuels and vehicles are obviously linked, in terms of energy use for transportation. The use of diesel engines and electric-drive vehicles, for example, dictates certain fuel choices. However, an increased variety of liquid fuels that can substitute for gasoline or diesel is emerging. In particular, biofuels, of which there are several different types, have received considerable attention.

Biofuels. A biofuel is any kind of fuel derived from organic material (biomass), ranging from plant crops to agricultural waste. Biomass can be burned directly without processing, and the heat from combustion is used to generate electricity. Biomass can also be processed into liquid fuel (e.g., biodiesel or ethanol). Not all biofuels are produced in the same way, and each has specific social, economic, and environmental impacts.

Starch ethanol. Most of the ethanol currently produced in the United States is derived from starches and sugars in the fruits and seeds of plants. The majority of the material in any given plant is fibrous cellulose, or lignin, which cannot be converted to an energy source through this process. In the United States, 95 percent of starch-based ethanol is produced from corn, a practice that has recently been strongly criticized for several reasons. Ethanol from corn has a relatively low net-energy ratio, which is a comparison of the nonrenewable energy put into production (cultivation, fertilizer, transportation, processing) with the amount of energy contained in the end product. Corn requires an enormous amount of energy, water, land, and fertilizer to grow, although in recent years ethanol production has increased its efficiency. It is nevertheless a relatively inefficient fuel source (USDA 2002).

Growing corn for fuel is problematic for other reasons as well. The rise in ethanol production has been blamed by some for driving up food costs due to the use of corn for ethanol and the conversion of acreage from other food crops to corn; others blame the increase in food costs on rising energy costs in general, since ethanol production has been under way for years. Growing corn for fuel does have several negative environmental consequences. Corn is generally grown using large amounts of fertilizer, herbicides, and pesticides, which can pollute water sources and cause algal blooms that deplete dissolved-oxygen content in water, killing fish and other aquatic species. Finally, fast-growing annuals like corn sequester very little carbon. (On sequestration, see Chapter 11.)

Cellulosic ethanol. Instead of using only the starch-heavy parts of a plant for fuel, cellulosic ethanol is made by breaking down the fibrous cellulose in plants. This process uses enzymes to break down the tough cellulose in plant cell walls into sugars. Microbes then convert the sugars through fermentation into liquid ethanol. The net energy ratio of cellulosic energy is estimated to be at least twice that of starch ethanol. (See Schmer et al. 2008.)

Unlike the feedstocks for starch ethanol, those for cellulosic ethanol require less energy and fewer chemical inputs, are less expensive, and are less detrimental to the environment. These include waste from agriculture and forestry (wood chips, stalks, husks, straw) that is often burned, emitting GHGs without benefit. Cellulosic ethanol can be made from perennial crops such as switchgrass and miscanthus, as well as from trees or shrubs. These plants, with their extensive root systems, can sequester large amounts of carbon, provide habitats for wildlife, effectively filter rainwater, and require little or no fertilization to yield usable biomass.

Currently, cellulosic ethanol is not cost-competitive. However, the U.S. Department of Energy is investing up to $385 million for six biorefinery projects. When fully operational, the biorefineries are expected to produce more than 130 million gallons of cellulosic ethanol per year, which is projected to make cellulosic ethanol cost-competitive with gasoline by 2012 (DOE 2007a).

Biodiesel. Biodiesel is a biofuel produced from biological-based oils. Most biodiesel is made from soybean oil; however, canola oil, sunflower oil, recycled cooking oils, and animal fats are also used. Growing terrestrial crops for biodiesel poses many of the same issues that ethanol production does. Biodiesel production can have greater or lesser energy ratios, depending on how the biofuel is produced, and biodiesel production can also compete for food and land resources. For example, increasing demand for biodiesel feedstocks has been linked to land conversions from tropical rainforest to oil palm plantations in Brazil and southeast Asia (Hill et al. 2006). However, a lesser-known source of feedstock, microalgae, can produce more gallons of biodiesel per acre than any terrestrial crop.

Biodiesel from algae. Given foreseeable yield rates and finite land resources, conventional crops have limitations in meeting our need for sustainable liquid fuels. As gas prices rise and technology continues to improve, however, an alternative energy crop, microalgae, is becoming increasingly viable.

Algae are microscopic organisms that live in a variety of aquatic ecosystems and produce energy through photosynthesis. Algae are extremely fast growing and produce large amounts of lipids, which help them maintain buoyancy near the water's surface, where they can access sunlight.

The U.S. Department of Energy has identified many species of algae that are more than 50 percent oil by dry weight; some strains can be genetically altered to have even higher lipid content. When the DOE studied the potential of algae for biofuel production under the Aquatic Species Program, its studies (like many others) showed that algae can produce far more biodiesel per acre of land used than any other terrestrial crop.

ECONOMIC BENEFITS OF LOCAL ENERGY

The average U.S. state spends a total of roughly $4,000 per person on energy (for all sectors—residential, commercial, industrial, transportation; EIA 2009a). For example, in 2007, Missouri, with a population just under six million, spent approximately $23 billion on all forms of energy. Past studies have estimated that for many states at least 50 percent or more of these energy dollars are sent out of state and not reinvested or recycled within the local or state economy—estimates have ranged upward of 90 percent for some states. Analysis by the American Council for an Energy-Efficient Economy indicates that energy efficiency and local renewable energy can increase retention of local energy dollars by 30 to 50 percent.

TABLE 2.1. BIOFUEL PRODUCTION PER ACRE

Crop	Gallons of Biofuel
Corn	18
Soybean	48
Cottonseed	120
Palm Oil	600
Algae	10,000

Algae are grown most successfully in closed bioreactors. This limits water waste from evaporation and keeps salinity levels in balance, maximizing oil production. To photosynthesize, microalgae need carbon dioxide, which can be diverted from power-plant flues.

After the lipids are pressed out of algae and the water is recycled for the next batch, the substance left over is a valuable feedstock for other applications. It can be dried and burned in electricity generators or used as feedstock for high-quality fertilizers and animal feed. In addition, methane can be produced through the anaerobic digestion of the leftover organic material. After digestion, it makes an excellent soil amendment. Some strains of microalgae produce hydrogen gas when deprived of sulfur and other minerals; research is being conducted on the production of hydrogen for fuel from algae (NREL 2008).

Renewable electricity for transportation. Electricity from renewable sources may also be an important energy source for transportation in the future. Electric motors have significant advantages in converting stored energy into mechanical power—they are typically twice as efficient as liquid-fueled combustion engines. Some studies have calculated that, in many cases, it would be more efficient to use biomass to generate electricity for transportation than to convert biomass into liquid fuel (Hamilton 2009). Electric motors in automotive vehicles can be powered by batteries, fuel cells, or other storage mechanisms. Transportation technologies are discussed further in Chapter 8.

LOCAL ENERGY STRATEGIES

Local communities are faced with many of the same energy issues that are before the United States as a whole. Local energy demand is likely to be trending upward. Concerns about sources, prices, and reliability of energy supplies may be intensifying, along with interest in developing new alternatives to meeting long-term local energy needs.

Choices made at the local level will increasingly shape how energy needs are met and how local and regional energy security, as well as national energy security, is achieved. Approaches will vary, but there are common challenges and opportunities that all communities will face.

Energy use and energy efficiencies vary widely from region to region, state to state, town to town, business to business, house to house, and person to person, due to historical, social, political, economic, and physical differences. And yet these differences are partly illusory; we live in a dynamic and interconnected national and global economy in which it is virtually impossible to define the boundaries of energy use precisely. However, every community can use energy more efficiently and find new ways to meet its local energy needs in ways that are more reliable, affordable, and sustainable over the near and long terms.

A local energy strategy can help build resilience to outside supply and price shocks, as well as recycle energy dollars within the local economy by making better use of local and regional energy resources. Progress on a local energy strategy can be measured either by the amount of energy use that is saved through efficiency or by the amount of imported energy that is displaced with local, renewable energy. By establishing a baseline of current and projected consumption rates for different types of energy, communities can begin to track how investments in energy efficiency and local energy are enhancing their energy security.

A local energy strategy and plan is a good way for a community or region to assess the potential for different energy options and to anticipate future energy needs and use. See Chapter 9 for a case study of an energy plan.

CHAPTER 3

Climate Change

 Climate change and energy issues are closely interrelated. Energy usage, both type and amount, is influencing the extent and rate of climate change, while climate change is influencing weather conditions that affect energy use.

Planners need to have a basic understanding of why scientists believe the earth's climate patterns are changing and what impacts such changes may have both globally and in different regions of the United States. This chapter addresses the causes and effects of climate change, while subsequent chapters discuss opportunities for mitigating and adapting to climate change. Here we provide a general treatment of climate science utilizing data and graphics from the Intergovernmental Panel on Climate Change (IPCC, an international collaboration of climate scientists sponsored by the United Nations and the World Meteorological Organization) and the U.S. Global Climate Change Research Program, among other sources. In addition, this chapter reviews both national and regional impacts that may result from climate change, including higher temperatures, sea-level rise, precipitation variability, and extreme weather events.

MITIGATION AND ADAPTATION

Responses to climate change can be put into one of two categories: adaptation or mitigation. Adaptive responses are efforts that address the impacts of climate change. For example, drought, intense precipitation, sea-level rise, species migration, and heat waves might be addressed by adaptive measures such as water resource management, stormwater control, coastal retreat, open space acquisition, and the provision of air-conditioned shelters for at-risk populations. By contrast, mitigative strategies address the "cause" of human-induced climate change. These might include attempts to reduce carbon dioxide (CO_2) and other greenhouse gas emissions through reductions in vehicle-miles traveled, the adoption of green building techniques, and reforestation.

In planning literature and in practice, more attention has been paid to mitigation than to adaptation. This is not surprising, since planners are trained to recommend community actions that reduce future adverse impacts. Much of what planners currently work on, such as discouraging sprawl development, encouraging green building, and facilitating multimodal transportation, are in themselves mitigation measures.

However, leading climate experts predict that some impacts from climate change will still occur. This is because CO_2 already in the atmosphere will remain there for many decades, contributing to further warming. It will take time to reduce GHG emissions to levels that leading scientists recommend. Global temperatures are predicted to rise throughout the century. Therefore, communities will need to prepare for and adapt to the anticipated impacts of climate change in their region.

Planners should apply their unique skill set and strategic position to address both the mitigation and the adaptation components of climate change response. As the IPCC (2007a) puts it:

> There is high confidence that neither adaptation nor mitigation alone can avoid all climate change impacts. Adaptation is necessary both in the short term and longer term to address impacts resulting from the warming that would occur even for the lowest stabilization scenarios assessed. There are barriers, limits and costs that are not fully understood. Adaptation and mitigation can complement each other and together can significantly reduce the risks of climate change.

WEATHER VERSUS CLIMATE

It is important to define what we mean by climate and climate change before discussing why climate change is happening and what effects may occur as a result. A common point of confusion lies in the distinction between climate and weather.

A simple definition of weather is "the state of the atmosphere at any particular time and location" (Buckley et al. 2004, p. 18)—which implies that weather is variable in both time and place. Weather is the end result of a complex global interaction between warmer and cooler air masses. The circulation of air molecules around the earth is also influenced by the rotation of the planet, the angle of the earth's orientation to the sun, and ocean currents, among myriad other factors. Given all these variables, the relative accuracy of present-day weather forecasting testifies to the ability of scientists to analyze complex data. Even though the occasional picnic is unexpectedly rained out or snow falls 50 miles north of where it was projected to, the improved accuracy of meteorologists' forecasts is one of the more significant scientific accomplishments of the past century.

Climate, on the other hand, is characterized by long-term patterns and cycles of weather, not by weather conditions on any one day. A nontechnical definition of climate would be the seasonal variation of the weather that is relatively constant over a significant period of time across a large region. The IPCC elaborates on that definition.

> Climate in a narrow sense is usually defined as the "average weather" or more rigorously as the statistical description in terms of the mean and variability of relevant quantities over a period of time ranging from months to thousands or millions of years. The classical period is 30 years, as defined by the World Meteorological Organization (WMO). These relevant quantities are most often surface variables such as temperature, precipitation, and wind. Climate in a wider sense is the state, including a statistical description, of the climate system.

What Is "Climate Change"?

Since climate is the average weather patterns of a geographic region, climate change indicates a significant measurable shift in those patterns. Climate indicators such as annual precipitation, daily high and low temperatures throughout the year, timing and length of seasons, frequency and severity of extreme weather events, and other key factors might all show substantial variations as the climate changes.

The IPCC defines climate change as a function of measurement:

> [A] change in the state of the climate that can be identified (e.g. using statistical tests) by changes in the mean and/or the variability of its properties, and that persists for an extended period, typically decades or longer. It refers to any change in climate over time, whether due to natural variability or as a result of human activity.

Note that there is a difference between climate change and weather variability. Over time, constancy in weather patterns during different times of the year creates seasons, each with specific associated weather characteristics. In most of the United States, we associate hot temperatures with summer; when there are cool summertime temperatures, we say the weather is "unseasonable." A region may even experience a series of unusually cool or warm, rainy or dry seasons. Although weather can vary significantly from year to year, a couple of hot summers or cold winters do not necessarily mean that a climate shift is occurring. However, when such anomalies become more commonplace over a longer period of time and other climate indicators show a sustained altered pattern, then the term "climate change" begins to apply.

What Causes Climate to Change?

The earth's climate has been shown to vary widely over time for natural reasons. The earth has experienced a wide range of climatic circumstances, from intense heat to intense cold, over the 4.5 billion years of its existence. Figure 3.1 shows these fluctuations and provides a good general illustration of global temperature changes.

Figure 3.1. Global temperature changes

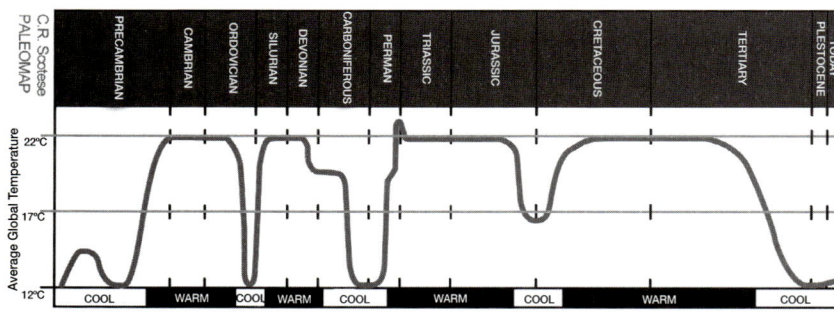

Source: Christopher R. Scotese, Paleomap Project, 2002

These climate changes were the result of a variety of factors, including volcanic activity, the intensity of solar radiation reaching the earth, major geologic events (such as the rise of the Himalayas), meteor impacts, and periodic fluctuations in the earth's orbit around the sun. These climate changes have been significant in relative terms, ranging from the incredible cold of the Cryogenian Period (850 to 630 million years ago, when the earth's surface may have frozen over entirely) to the warmth of the Jurassic Period (commonly thought of as the time of the dinosaurs, 206 to 144 million years ago, when tropical conditions were present as far north as Alaska). In general, however, the world's temperature has been reasonably constant compared to the extremes seen on other planets in our solar system. This constancy is a result of moderating influences from natural "regulators" like the greenhouse effect and the carbon cycle.

The greenhouse effect and the carbon cycle have played—and continue to play—critical roles in moderating temperatures on Earth. Simply put, the greenhouse effect keeps the warmth from solar radiation that reaches the Earth from escaping into space, creating an atmospheric "blanket." The carbon cycle is a complex process in which various forms of carbon are sequestered or stored in geologic formations, such as the White Cliffs of Dover and the limestone underlying much of the state of Florida, and in plants and animals (biomass). This stored carbon is released into the atmosphere through plant and animal respiration, decomposition of biomass, forest fires, and volcanic activity. This release is a significant component of the greenhouse effect, although other greenhouse gases like methane and water vapor also play a part. Natural exchanges such as uptake by the ocean's surface and photosynthesis return atmospheric carbon to the earth's surface, where it is again placed in geologic or biomass storage.

Less frequent volcanic activity over the past half-billion years has slowly reduced CO_2 concentrations in our atmosphere, allowing the earth to cool. Combined with the orbital fluctuations that affect the amount of sunlight that reaches the earth's surface, this pattern has led to a series of ice ages in which long periods of glacial activity have been interspersed with shorter periods of global warming. These are illustrated in the temperature valleys and peaks in Figure 3.2; the close correlation between temperature and CO_2 will be further discussed below.

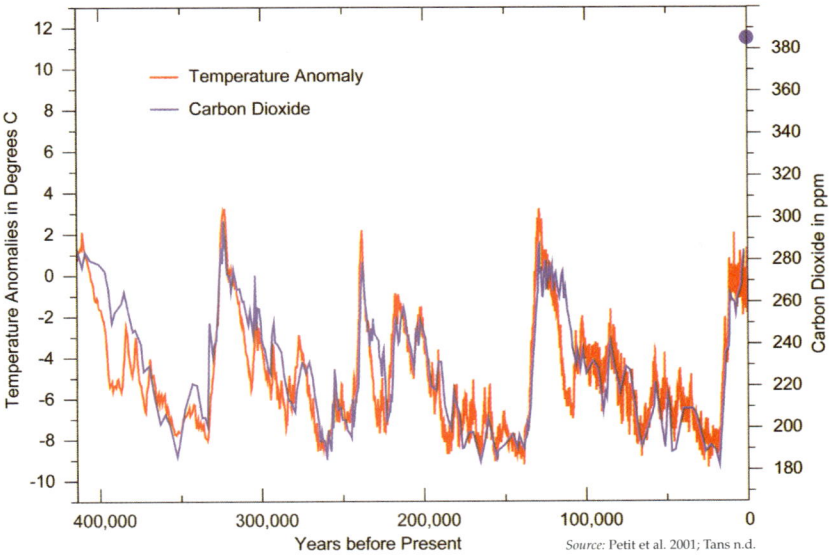

Figure 3.2. Temperature/CO_2 correlation

The advance and retreat of glaciers have shaped our landforms, especially in the northern hemisphere, and have resulted in dramatic changes in global sea levels as our planet's water was converted to ice and then returned to the oceans as the glaciers melted. We are currently in what is referred to as an "interglacial" period marked by retreating glaciers and less extensive sea ice.

Most scientists feel that climate is now being influenced by human activities, in addition to these natural fluctuations. This phenomenon is described in the following section.

Recent Climate Change

Changes in the earth's climate observed over the past 200 years fall outside the variability that scientists would expect from natural factors. What could be causing the climate to change to this extent?

For the past 800,000 years, the amount of CO_2 in the atmosphere has stayed relatively constant, occurring in a band ranging from approximately 200 to 300 parts per million (ppm), with the lower CO_2 concentrations occurring during periods of colder climate. This relative constancy has been due in part to the stor-

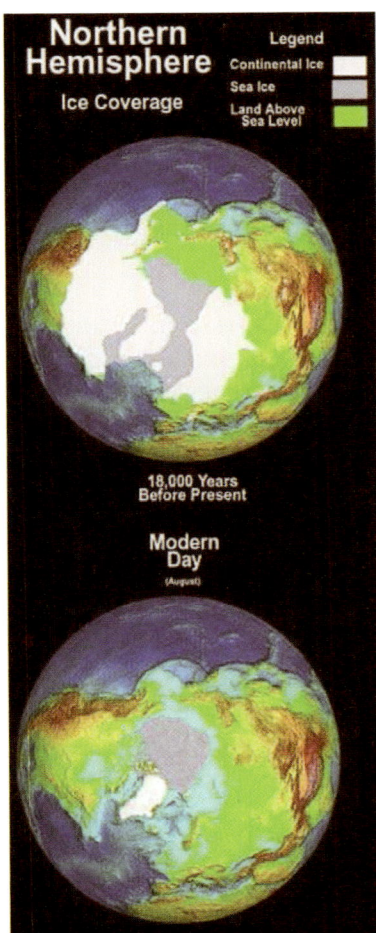

Figure 3.3. Sea levels during the last ice age

age or "sequestering" of CO_2 by plants and animals in living and fossilized forms. Fossilized forms of carbon include oil, peat, natural gas, limestone, and coal.

Starting in the mid-1700s, the Industrial Revolution and modern agricultural practices began to alter this trend. Fossilized forms of carbon were burned to create the energy needed for industrial processes, maintenance of indoor environments (heating and cooling), and transportation. Natural ecosystems were modified by agricultural practices, including forestry, and by development. The growth in human population that has been supported by these practices now depends upon them for its maintenance and continued expansion, creating a self-perpetuating cycle. As a result, CO_2 concentration in the atmosphere has increased by more than 30 percent, from 275 ppm in the early 1700s to about 385 ppm today. Projections of atmospheric CO_2 concentration for the year 2100 range from approximately 550 ppm to over 900 ppm, depending on the extent of GHG emissions that occur.

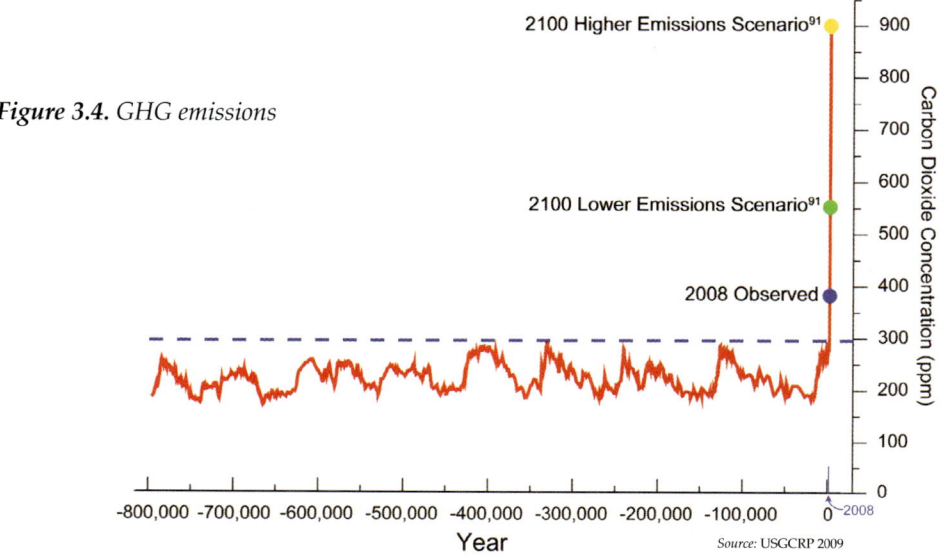

Figure 3.4. GHG emissions

Human activity, from the burning of fossil fuels to deforestation for agriculture and development, is creating what scientists call a "forcing" event. Instead of CO_2 concentrations trailing changes in global temperatures, they are now leading to planetary temperature change. The uniqueness of this circumstance makes predicting its consequences even more difficult, especially since there is considerable uncertainty regarding future economic development, population growth, and the success of human efforts to mitigate the amount of CO_2 being released into the atmosphere.

However, significant scientific resources have been devoted to identifying potential climate-change impacts based upon various assumptions or scenarios. This research indicates that, regardless of our current and future efforts to modify our activities that are effecting climate change, significant changes to our climate are inevitable because of the increase in CO_2 concentrations that has already occurred. The following section describes the impacts associated with a warming climate.

Global Climate-Change Impacts

We now face a period of relatively abrupt climate change. Natural systems developed over centuries will have to adjust to the effects of increased temperatures, changes in precipitation patterns, greater frequency of extreme weather events, and other manifestations of climate change on a global scale. Human systems will be similarly challenged to adapt.

Scientists believe present GHG levels already commit us to some future warming (1–2°F). IPCC climate models project an average global temperature rise of between 2° and 8°F by 2100. The IPCC further projects that these temperature increases will result in a global sea-level rise somewhere between 0.18 and 0.59 meters (0.6–1.9 feet) by 2100. This IPCC estimate is now considered conservative because it did not account for contributions from melting ice sheets such as those covering Greenland and Antarctica, due to uncertainties in measurement at the time of the estimate.

It is critical to a sound understanding of climate change to realize that regional effects may be substantially more extreme than global changes might suggest. This is because global temperature measurements include the relatively slow-to warm oceans and because global estimates for sea-level rise do not account for geologic subsidence or uplift of coastal areas. Consequently, average surface temperatures over land in areas such as Alaska could rise by nearly 20°F, and relative sea-level rise along the Gulf Coast, an area experiencing subsidence, could well be substantially more than two feet, according to the IPCC.

Climate Change Impacts in the United States

Climate change affected past civilizations through drought and adverse temperatures. In more recent times, technological advances have allowed people to inhabit perhaps otherwise unfriendly climates. Air-conditioning has allowed hot and humid Florida and hot and dry Arizona to enjoy significant population growth over the past 50 years. Indoor heating allows Alaska to be inhabited by a substantial and growing number of people year-round.

We are learning that the post–World War II expansion of population in the United States occurred during a time of extremely favorable climate for such an expansion. Abundant water and electricity allowed people to inhabit previously hostile environments. Abundant and inexpensive fossil fuels resulted in dramatic changes in development patterns and supporting transportation networks. Development in the United States moved from a predominantly urban pattern to a predominantly suburban one over the course of just a few decades. The nation's population center moved both southward and westward. These physical changes in the developed landscape were facilitated by the very factors that are so influential in manifesting today's climate change: consumption of fossil fuels and conversion of forested areas and plains into developments and farms.

In the western United States, the region's climate is expected to revert to a drier precipitation pattern, so existing water supplies will be reduced. If the existing development there is to be sustained, it will require extraordinary engineering, conservation, and intergovernmental cooperation measures. Higher summertime temperatures and longer warm seasons are anticipated to increase cooling costs across the nation but especially in the Sun Belt. More frequent heat waves could threaten an increasingly vulnerable population across the United States, especially in northern states, where air-conditioning is not as commonplace, but heat wave–induced electrical blackouts and brownouts could affect every region. Coastal areas are expected to be affected by sea-level rise.

The U.S. Global Climate Change Research Program recently released *Global Climate Change Impacts in the United States*, which provides an accessible assessment of climate change–induced impacts in the country as a whole and in specific regions. Using information from this publication, the following sections contain information about specific national climate-change impacts, including natural climate-change indicators such as surface temperatures, ocean temperatures and sea-level rise, precipitation, and extreme storms and natural disasters.

Surface temperatures over land. The gradual increase in average global temperatures can manifest itself in more dramatic temperature increases in specific regions. For example, the U.S. average temperature has risen more than 2°F since 1950, more than double the global increase over the past century. There will continue to be considerable regional variation in projected temperatures, depending upon both location and emissions scenarios, as illustrated in Figure 3.5. On average, end-of-century temperatures are expected to be 4° to 11°F warmer, depending on region. Northern and central regions will see temperature increases regardless of which emissions scenario prevails. Alaska, in particular, will be significantly affected by rising temperatures.

Figure 3.5. Projected temperature changes (°F), from 1961–1979 baseline

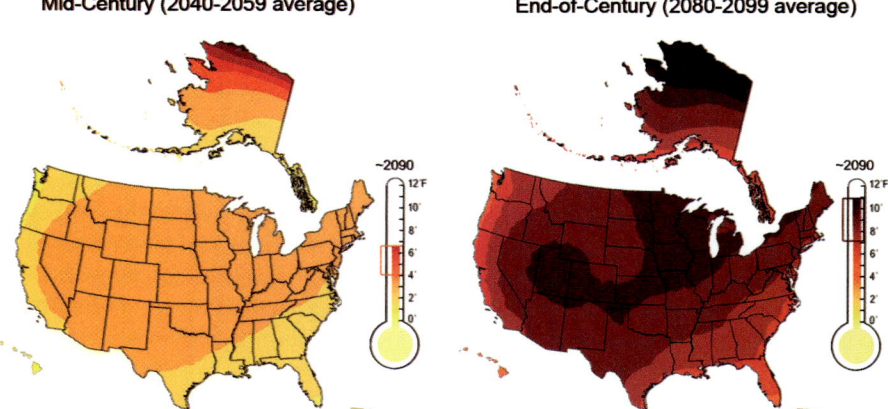

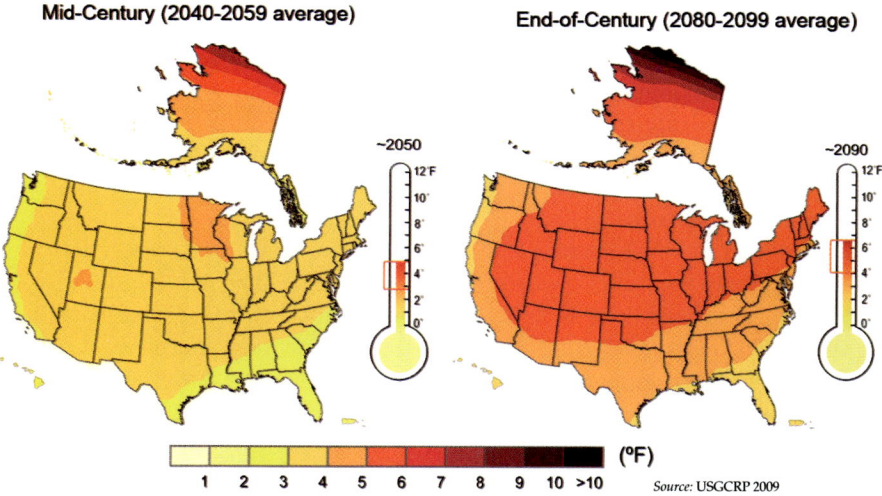

Another way to measure increases in temperature is by the number of days with highs above 90°F. Figure 3.6 shows an estimate of the number of such days per year in various areas of the continental United States under two emissions scenarios, using recorded observations from the recent past. Under the high emissions scenario, daily high temperatures by 2099 are anticipated to exceed 90°F in Florida, much of south central Texas, and southern Arizona and New Mexico on more than 180 days per year.

Recent Past (1961-1971 Average)

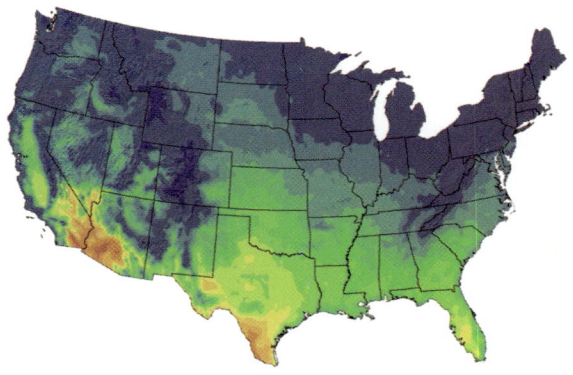

Figure 3.6. Number of days with highs above 90°F, in different emissions scenarios

End-of-Century (2080-2099 Average) under Lower Emissions Scenario

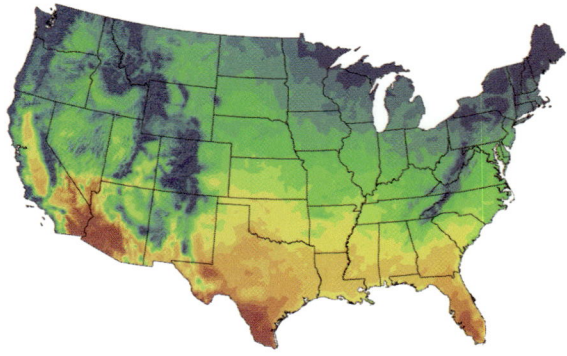

End-of-Century (2080-2099 Average) under Higher Emissions Scenario

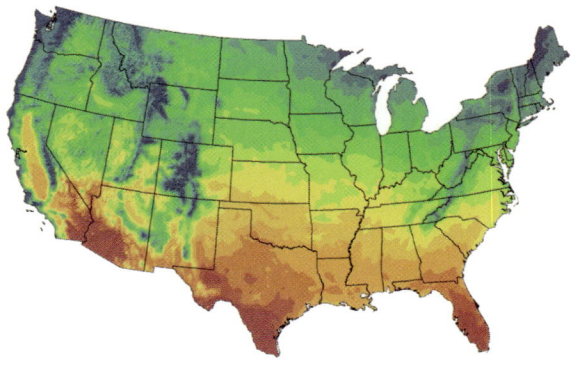

Number of Days per Year

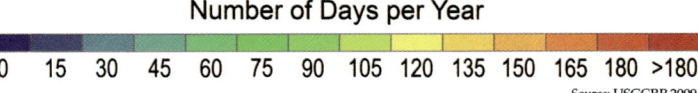

0 15 30 45 60 75 90 105 120 135 150 165 180 >180

Source: USGCRP 2009

One of the most troubling effects from these temperature increases involves heat waves. Figure 3.7 illustrates the historical trend of very hot (meaning in the 90th percentile or above) July days in the United States from 1950 to 2004. The graph shows an increasing temperature trend for both daily maximum and minimum temperatures, with the daily minimum temperature increasing more rapidly than the maximum. This means that it has not been cooling off as much at night, creating the potential for longer-lasting heat waves and greater potential for heat-related mortality.

Figure 3.7. Very hot days

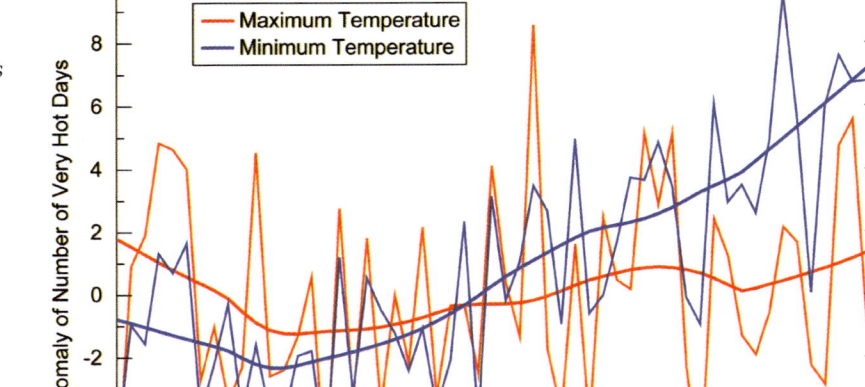

Heat waves, also known as excessive heat events (EHEs), are among the most lethal type of extreme weather event. They will significantly increase in frequency and length as a result of climate change. In 2004, researchers analyzed climate models based on scenarios from the IPCC for the State of California and found that by 2070–2099, extreme high temperatures (defined as about 90°F) would increase from a historical norm of 5 percent of total days per year to 12 to 19 percent under a low emissions scenario and to 20 to 30 percent under a high emissions scenario, or up to six times the historical level (Hayhoe et al. 2004).

The increase in heat waves will have consequent mortality effects (for more information on EHEs, see U.S. EPA, Office of Atmospheric Programs 2006). A 2008 study examined the potential effects of an EHE, analogous to the European heat wave of 2003 that killed more than 35,000 people, in five large cities in the United States (Kalkstein et al. 2008). The projection indicated that 4,906 deaths would have occurred, 3,253 of them in New York City alone. Many U.S. cities are relatively unprepared for heat waves even though the mortality in these five cities for the hottest summer between 1961 and 1995 totaled 2,716, significantly more than the death toll from Hurricane Katrina. Two heat-wave events in the 1980s killed nearly 20,000 people in the United States.

Mortality during heat waves is today often associated with loss of electric power during brownouts and blackouts. Electrical-service supply shortages are generally considered to be a universal problem; in the event of a shortage, all population sectors are assumed to be affected equally. That is not entirely the case, however. Persons in high-crime areas may refuse to evacuate to air-conditioned shelters due to fears about theft or vandalism, while the elderly or other segments of the population may not have access to transportation to such shelters. Plant shutdowns can create exorbitantly high electricity costs since utilities are forced to purchase expensive peak power from other sources. These expenses are typically passed on to the customers and may cause lower-income persons to cut back on or preclude their use of electricity, especially air-conditioning, even when it is available. Public-safety service providers need to take these considerations into account in their planning and service delivery. Planners can help address these problems by locating concentrations of vulnerable populations and developing effective outreach and adaptation techniques to assist these people. (See Table 3.1.)

TABLE 3.1. SOME POSSIBLE EHE RESPONSES

Planning Activity	Possible Responses	Comments
Delivery of social services	• EHE awareness programs. • Coordination of EHE shelter operations. • Medical assessment and treatment programs. • Vulnerable population database. • Distribution of light-colored clothing. • Employer outreach.	Reaching vulnerable populations with information and services is critical prior to and during EHEs. Consuming alcohol and wearing dark clothing exacerbate heat effects. Employers need to understand the health consequences of strenuous outdoor work during EHEs.
Design of buildings and sites	• Design for natural ventilation. • Install awnings and overhangs. • Shade-providing landscaping. • Backup electrical systems/generators. • Use of heat-dissipating materials & colors. • Retention of mature landscaping for shade. • Energy-efficient design.	Natural ventilation, especially in multifamily buildings, helps counter EHE effects. Energy-efficient design makes it more likely for lower-income persons to use their air conditioners. Natural landscaping lessens the urban heat island effect.
Neighborhood and comprehensive planning	• Neighborhood-based shelters. • Vulnerable population database helps direct capital improvement programs.	Advance planning for EHEs in neighborhood and community design is helpful.
Affordable housing	See design of buildings and sites, above	
Recreational programs and amenities	• Provision of shade in parks and playgrounds. • Changing timing/programming of outdoor recreation programs.	Timing and programming (e.g., keeping pools open) offer reduced exposure to and relief from heat.
Public safety planning	• Early warning alerts. • EHE shelter management operations. • Coordination with electric utilities during EHEs. • EHE awareness programs. • Vulnerable population database. • Increased law enforcement in high-crime areas during EHEs.	Early warning alerts from public safety officials get more attention. Coordination with utilities is critical. Knowing the location of vulnerable populations enhances response. People in high crime areas may not open windows even during EHEs.
Utility system management, especially electricity and water supply	• No scheduled maintenance or shutdowns during EHEs. • Coordination with public safety officials during EHEs.	Continuity of utility operations is critical to counter the effects of EHEs.

Ocean temperatures and sea-level rise. Perhaps even more important than increases in surface temperatures over land is the rising temperature of the world's oceans and its consequent effect on sea level. Sea-level rise is a function of four factors. The first and perhaps most obvious one is the general increase in the amount of water in the oceans resulting from runoff from melting snowpack and glaciers. The second factor is "thermal expansion" of the oceans; warm water contains a greater volume than cold water, so higher water temperatures mean greater oceanic water volume. The third factor is the contribution of melting ice sheets, such as those that cover Greenland and Antarctica. The fourth factor is geologic: whether the land affected by the general increase in the amount of the ocean water is uplifting, subsiding, or stable. Sea-level rise will be much more pronounced in areas experiencing subsidence, such as the Gulf Coast, than in areas experiencing uplift, such as portions of Alaska.

Perhaps even more important than increases in surface temperatures over land is the rising temperature of the world's oceans and its consequent effect on sea level.

During the 20th century, coastlines in the United States and throughout the world have been extensively developed. This development has been supported by infrastructure, including water- and sewage-treatment plants, roads, airports, and even hospitals that have been located in flood-prone areas. Rising sea levels could threaten this development and infrastructure by increasing their vulnerability to tidal flooding, saltwater intrusion of freshwater aquifers, and the consequences of extreme weather events such as hurricane storm surge. Figure 3.8 illustrates the effect of sea-level rise on two northwestern cities, Olympia and Seattle, Washington.

Figure 3.8. Local sea-level rise impacts

Precipitation changes and drought. *Global Climate Change Impacts in the United States* states that "projections of future precipitation generally indicate that northern areas will become wetter, and southern areas, particularly in the West, will become drier." This is depicted in Figure 3.9, which illustrates anticipated future seasonal changes in North American precipitation for the period 2080–2099.

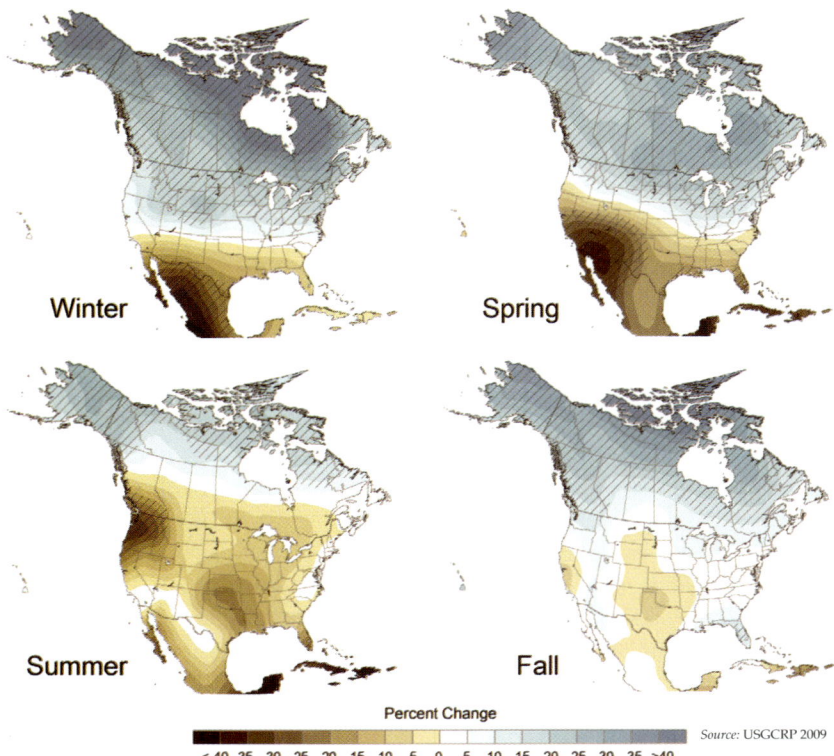

Figure 3.9. Anticipated precipitation changes

While these precipitation changes have a variety of effects, including flooding (discussed below), perhaps the most serious involves drought. Drought conditions may become more frequent as a result of climate change, especially in the western United States. The largest possible impact would be on reductions in the amount of water available for agriculture and direct human consumption (potable and industrial uses), although habitat loss and species decline and extinction are also concerns. A California study projects declines in the Sierra snowpack of 30 to 90 percent by the latter part of the century, depending on which emissions scenario emerges, "with cascading impacts on runoff and streamflow that, combined with projected modest declines in winter precipitation, could fundamentally disrupt California's water rights system" (Hayhoe et al. 2004). Such environmental conditions could have enormous consequences for the future of the region.

Decreases in water availability would require that state and local governments throughout the region work together to mitigate and adapt to the projected changes. Some steps are already being taken. For example, a major cooperative agreement was reached between the seven basin states of the Colorado River in 2007 regarding management of the river. Conflicts among water user types will require resolution; for example, agriculture, a major user of water in the West, will find itself in increasing conflict with communities that need water for their citizens and industries.

Outside of the West, drought could also present a problem. Recent extreme drought conditions in the southeast have prompted extensive conservation measures in many locations and litigation in others.

The effects of drought conditions on public safety operations include insufficient potable water supplies and the potential for brownouts and blackouts as hydroelectric and nuclear power generation is affected by low water flows or decreased availability of water for cooling.

Extreme Storms and Natural Disasters

Extreme precipitation and flooding. Heavy rainfall events and rapid melting of snow and snowpacks are typical triggers of flooding. In some cases, flooding can be exacerbated by extremely dry soils created by drought, especially in areas of steeper slopes, which may lack the ability to absorb precipitation at normal rates.

Climate change is projected to create more heavy precipitation events, resulting in the potential for greater flooding, erosion, landslides, mudslides, avalanches, snow-load damage, and other precipitation-related impacts. Some regions may see increases in total annual precipitation, while other areas are projected to receive less rain and snow on average.

Communities which experience significant winter snowfalls may want to evaluate their emergency procedures for addressing roof collapse problems created by excessive snow loading. (Such snowfalls may also be "wetter"—and therefore heavier—as a result of rising temperatures.) The northeast experienced numerous roof failures during the record snowfalls of the winter of 2007–2008. Due to rising temperatures, ice storms could also increase in range and frequency.

Figure 3.10. Roof collapse

© iStockphoto.com/Tammy Bryngelson

Building codes and development standards must take these types of events into account in order to adapt to climate change effects. For example, flood elevation maps must be reevaluated and cooler-climate snow-load code standards must be revised in light of climate change effects.

Figure 3.11 illustrates projected changes in annual runoff by 2040–2060. Extremely dry areas in the southwest are projected to exhibit reductions in runoff, while other areas will experience significant increases, especially in the portions of the Midwest and in Alaska. These are annual projections and do not reflect flooding potential from specific events.

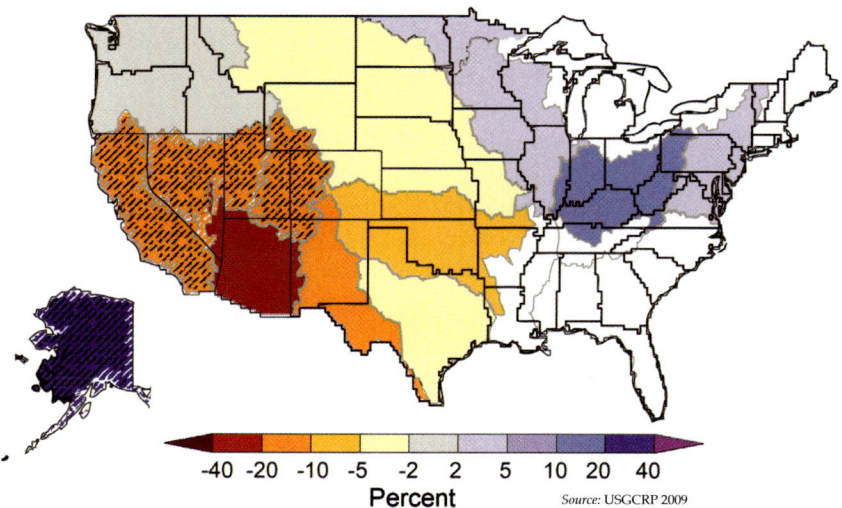

Figure 3.11. Projected changes in annual runoff

Projected increases in precipitation and flooding due to climate change have the potential to affect public health. Water-supply contamination due to storms and flooding could present a serious risk. Treatment facilities can be damaged or inundated by storm surge or floods. Underground and surface sources can become contaminated. Pollutants and diseases from upstream sources may affect the quality of water treated at treatment plants, even those which are not otherwise damaged.

In particular, joint stormwater and sewage-treatment systems (also known as combined wastewater systems) can create significant health problems in flood conditions. These systems are designed to divert excess wastewater into natural systems during floods. This creates the potential for contamination of downstream water supplies with pollutants and waterborne diseases. Such systems are concentrated in locations that will likely be more prone to flooding in the future, due to increased intensity of precipitation. (See Figure 7.9, page 83.)

Potential contaminants include *Cryptosporidium* (a protozoa that causes intestinal illnesses), *E. coli* (a bacterium that also causes intestinal illnesses), and various pollutants (such as pesticides and petroleum products). Monitoring these contaminants and, where possible, treating them is necessary to maintain a healthy water supply. Contaminated potable water supplies create a variety of health problems, including serious intestinal illnesses; until proper treatment capabilities are reestablished, alternative potable water sources must be provided.

Hurricanes, Tropical Storms, and Strong Nontropical Storms

Scientific modeling and projections are currently inconclusive as to whether hurricanes and tropical storms will increase in frequency due to climate change. However, there is a likelihood that the tropical events that do occur will be of greater intensity since tropical storm systems are extremely dependent upon warm surface waters for development and intensification. A general increase in the intensity of tropical systems may occur when water temperatures are increased by global warming.

Strong storms may include hurricanes, tropical storms, and northeasters that affect coastal areas in the eastern and southern United States and on tropical islands, as well as nontropical storms that affect all regions of the country. Higher temperatures from climate change may increase the intensity of northeasters and other nontropical storms. Meteorologists have observed many cases where tropical systems intensify just prior to landfall because they encounter warmer waters in the shallower areas near land.

Any of these storm types has the potential for torrential rain, with resulting local and regional flooding, flash flooding, erosion, mudslides and landslides, and infrastructure impacts, including sewer backups, dam and levee failure, and storm-sewer overflows. In areas that experience storm surge from hurricanes, tropical storms, and northeasters, substantial property and infrastructure damage can occur. Wind from these storms can also damage property and infrastructure and exacerbate tidal flooding problems. For example, in late 2008, an extreme precipitation event (both snow and rain) significantly affected portions of the Pacific Northwest, isolating the region from the transport of goods and severely limiting travel.

Public health risks to communities from these extreme weather events include:

- Loss of life
- Water supply contamination
- Shelter availability and event damage risks
- Postevent mental health problems
- Damage or restricted access to health-care infrastructure, such as hospitals and clinics
- Restricted access for emergency services
- Postevent risks from mold and contamination
- Limitations on travel, especially evacuation, and the transport of goods, including food supplies

Extreme weather events happen every year. Figure 3.12 illustrates the major weather disasters that occurred in the United States from 1980 to 2008. The locations and types of events provide an indication of regional vulnerability to extreme weather events. These events generally, but not always, reflect a severity associated with relative population density. This is particularly true for hurricane events.

Figure 3.13 illustrates the number of extreme storms and relative damage caused by these extreme events on an annual basis between 1980 and 2008. (Hurricane Katrina created an extraordinary spike in damages in 2005.) There seems to be an increasing frequency of extreme events, with more billion-dollar events occurring annually. It is not entirely clear that this relatively recent increase is directly tied to climate change, but growing evidence indicates that many types of extreme events may increase in frequency or intensity because of changes to our climate.

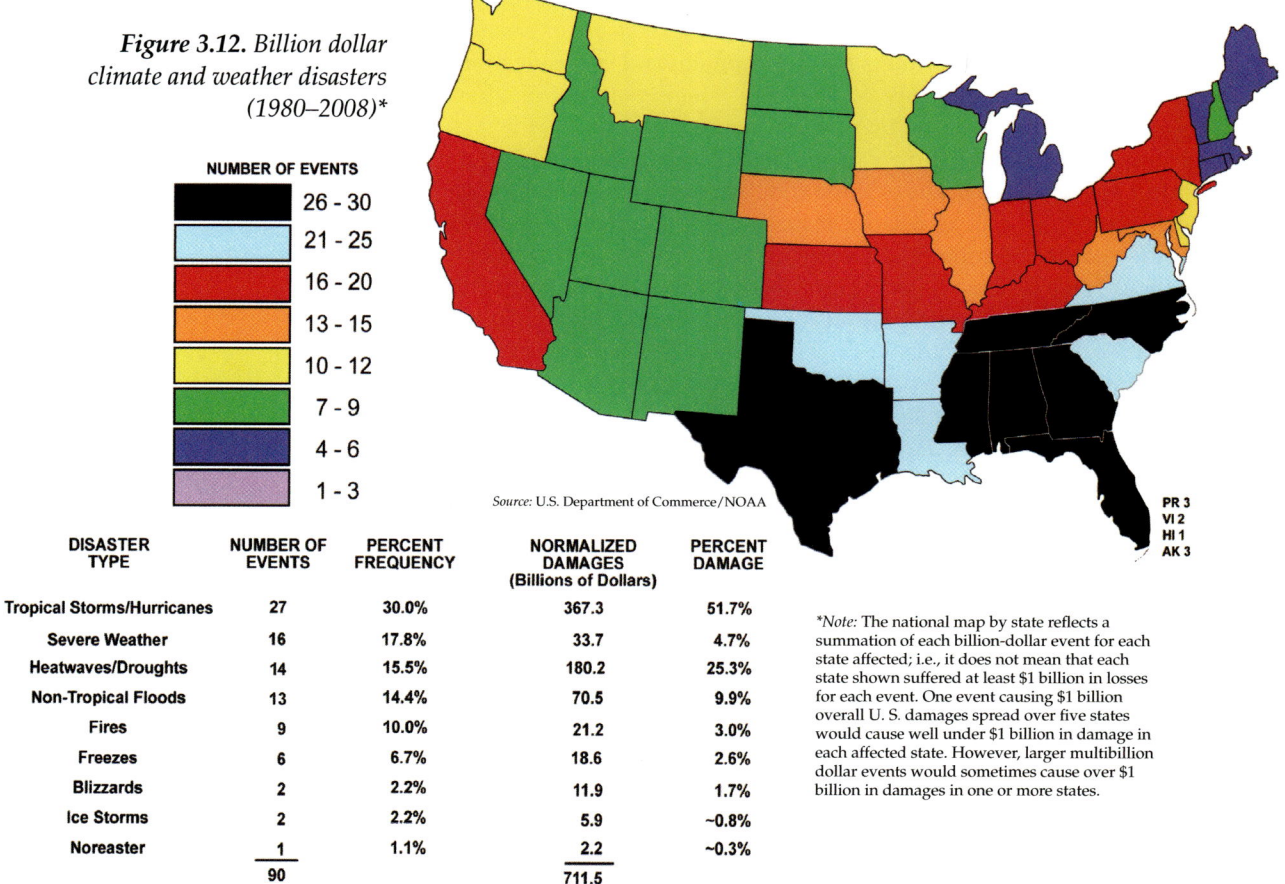

Figure 3.12. Billion dollar climate and weather disasters (1980–2008)*

DISASTER TYPE	NUMBER OF EVENTS	PERCENT FREQUENCY	NORMALIZED DAMAGES (Billions of Dollars)	PERCENT DAMAGE
Tropical Storms/Hurricanes	27	30.0%	367.3	51.7%
Severe Weather	16	17.8%	33.7	4.7%
Heatwaves/Droughts	14	15.5%	180.2	25.3%
Non-Tropical Floods	13	14.4%	70.5	9.9%
Fires	9	10.0%	21.2	3.0%
Freezes	6	6.7%	18.6	2.6%
Blizzards	2	2.2%	11.9	1.7%
Ice Storms	2	2.2%	5.9	~0.8%
Noreaster	1	1.1%	2.2	~0.3%
	90		711.5	

Note: The national map by state reflects a summation of each billion-dollar event for each state affected; i.e., it does not mean that each state shown suffered at least $1 billion in losses for each event. One event causing $1 billion overall U.S. damages spread over five states would cause well under $1 billion in damage in each affected state. However, larger multibillion dollar events would sometimes cause over $1 billion in damages in one or more states.

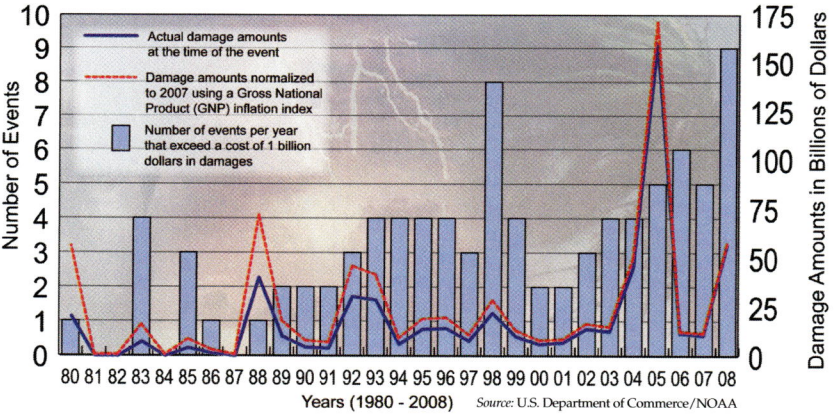

Figure 3.13. Billion-dollar U.S. weather disasters (1980–2008)

Wildfires

Wildfires are a natural occurrence in most forests. Forests have adapted to wildfires, which clear underbrush, promoting canopy-tree seed germination and growth and providing soils with beneficial minerals. However, as the development of the United States has progressed, natural wildfires have been suppressed to protect property and aesthetics, leading to a buildup of brush and other low-lying biomass. As a result, when a wildfire does occur, it has more fuel to burn. This causes hotter conditions that damage or kill trees, creating more fuel for future fires as well as other ecological damage such as erosion and habitat loss. According to the U.S. Geological Survey:

> Wildfires are a growing natural hazard in most regions of the United States, posing a threat to life and property, particularly where native ecosystems meet developed areas.

However, because fire is a natural (and often beneficial) process, fire suppression can lead to more severe fires due to the buildup of vegetation, which creates more fuel. In addition, the secondary effects of wildfires, including erosion, landslides, introduction of invasive species, and changes in water quality, are often more disastrous than the fire itself.

Drier conditions exacerbated by climate change, especially in the southwest, are anticipated to create situations that in the absence of enhanced management will be conducive to an increased frequency of wildfires. Currently, dry conditions in the western United States are resulting in major wildfire events, with millions of acres burned and thousands of homes destroyed.

Thanks to improved weather forecasting, special drought modeling, and real-time satellite imagery and analysis, wildfire conditions are typically anticipated and emergency services providers are alerted. Drought and wildfire services and Community Wildfire Protection Plans (see Chapter 11) can help address problems associated with wildfires.

REGIONAL CLIMATE-CHANGE IMPACTS IN THE UNITED STATES

As indicated, climate-change impacts are anticipated to vary by region across the United States. Planners must understand the regional effects of climate change in the communities they serve in order to for them to develop and implement effective mitigation and adaptation strategies.

Some of the more significant regional impacts identified in *Global Climate Change Impacts in the United States* (2009) include:

Northeast Region
- Shorter winters with fewer cold days and more precipitation; significant reductions in the winter snow season
- Under the higher emissions scenarios, 20 to 30 days in which the high temperature in cities exceeds 100°F; more frequent heat waves; and, on average, six additional weeks of summer conditions
- More frequent flooding as a result of sea-level rise and heavy precipitation events
- Economic effects including negative impacts on agricultural production, such as dairy, fruit, and maple syrup, reduced snow cover adversely affecting winter recreation, and a northward shift of lobster fisheries and diminution of Georges Bank cod fisheries

Southeast Region
- Heat-related stresses for people, plants, and animals
- Decreased water availability due to increased temperature and longer periods between rainfall events
- Sea-level rise and the potential for increased hurricane intensity, significantly affecting coastal areas and ecosystems

Midwest Region
- In the summertime, increased heat waves and reduced air quality
- A longer growing season, potentially generating increased crop yields, provided challenges such as heat waves, floods, and greater numbers and varieties of pests can be managed
- Increased volatility in precipitation, resulting in more frequent flood and drought conditions

- Significant reduction in Great Lakes water levels as a result of higher temperatures that promote greater evaporation, affecting shipping, infrastructure, water-based tourism/recreation, and ecosystems

Great Plains
- Negative impacts on region's water resources resulting from increased temperature and evaporation and frequency of drought
- Stresses on agriculture, ranching, and natural lands management resulting from changes in precipitation and higher temperatures
- Negative effects on key habitats and ecosystems, especially wetland systems

Southwest Region
- Increasing scarcity of water supply, requiring policy decisions to prioritize allocation among competing uses such as urban populations and agriculture
- Increased temperature, drought, and wildfire, significantly affecting ecosystems
- Negative effects on tourism/recreation industries, including reduced snowpack in ski-resort areas and unique ecosystem degradation

Northwest Region
- Declining snowpack negatively affecting regional water supplies
- Higher temperatures increasing risks to forestry from wildfires and insect pests
- Negative impacts on coastal areas resulting from sea-level rise
- Decreasing habitat for cold-water fish, such as salmon

Alaska
- Higher temperatures increasing risks to forestry from wildfires and insect pests
- Longer growing season and longer periods for outdoor tourism due to increasing temperatures
- Damages to infrastructure due to thawing permafrost
- Negative effects on coastal areas from loss of sea-ice buffers, increasing frequency of strong storms, and thawing permafrost

Pacific and Caribbean Islands
- Reduction in availability of freshwater supplies due to changing rainfall patterns, including reduced precipitation in the Caribbean region and contaminated groundwater from flooding in the Pacific islands; sea-level rise will threaten underground freshwater supplies
- Negative effects on marine ecosystems, creating problems for tourism and fisheries industries
- Greater frequency of coastal inundation resulting from sea-level rise and increased intensity of storms

Regional climate-change impacts like those identified above create an opportunity for adaptive action by planners. Table 3.2 lists specific populations' particular vulnerability to different types of climate change–induced public safety impacts. Spatial information about most of these population groups is available through the U.S. Census and other sources, allowing planners to create population vulnerability maps that can be used to identify areas with populations requiring particular types of services during heat waves, storms,

TABLE 3.2. VULNERABILITY OF CERTAIN POPULATION SUBGROUPS TO DIFFERENT TYPES OF CLIMATE CHANGE–INDUCED IMPACTS

Population	Heat Waves	Strong Storms	Flood	Drought	Wildfire	Comments
Persons over 65	✓	✓	✓	✓	✓	Less able to withstand stresses created by heat waves and droughts; greater potential mobility problems in avoiding storms, floods, and wildfires
Persons 14 and under	✓					More likely to be outdoors for recreation
Persons with disabilities or chronic illnesses	✓	✓	✓	✓	✓	Health and mobility issues can compromise the safety of this group
Linguistically isolated persons (non-English speakers or those for whom English is a second language)	✓	✓	✓	✓	✓	Public-safety service delivery to these populations can be complicated by: • Communications or cultural issues • A tendency to be more likely to be employed in outdoor vocations or outdoors for recreation • Homelessness • A refusal to open windows during heat waves or to evacuate due to crime concerns • Reduced electrical use if cost is driven up by drought effects
Socially isolated persons, including the homeless	✓	✓	✓	✓	✓	
Single adults with children	✓	✓	✓	✓	✓	
Transportation-challenged (no car or access to transit) persons		✓	✓	✓	✓	
Persons residing in high crime areas	✓	✓	✓	✓	✓	
Persons residing in mobile homes		✓	✓	✓		
Persons with below median incomes	✓	✓	✓	✓	✓	
Persons residing in substandard housing	✓	✓	✓	✓	✓	
Persons residing in multifamily structures	✓					Less opportunity for cross-ventilation

floods, droughts, and wildfires. Such maps can be overlaid with information about topographic and other geography-related vulnerabilities (such as low-lying areas, areas prone to storm surge, areas likely to experience particularly heavy snow accumulation, areas having compromised evacuation routes, and so forth) in order to establish priority locations for enhanced or population-specific public-safety service delivery.

SUMMARY

Forecasting our climate future is extremely complex and is highly dependent upon future behavior. As a consequence, climate change projections are linked to certain assumptions, including the degree to which GHG emissions are moderated.

These projections are expressed in ranges of impacts—temperature increases, variations in precipitation, and extent of sea-level rise, for example. Even if only the lower levels of these ranges ultimately manifest themselves, it is clear that planners will need to be prepared.

However, it is apparent from IPCC scenario modeling that changes in the levels of GHG emission that result from human population growth, economic development choices, development patterns, energy usage, agricultural practices, and other factors can result in significantly different futures on account of global temperature rise and consequent climate change. Efforts to mitigate the extent of global temperature rise through reductions in the amount of GHG emissions have the potential to moderate future climate-change effects. Chapter 4 explores this in more detail.

CHAPTER 4

Greenhouse Gases

 As discussed in Chapter 3, global climate changes observed over the past 100 years are believed to have been caused largely by increased concentrations of greenhouse gases in the atmosphere. Scientists believe the biggest contributors to these increased concentrations are human activities: the burning of fossil fuels for energy, clearing and burning of forests, agricultural practices, and certain industrial processes in particular. As noted, global warming and climate change are projected to continue to some extent as a result of past GHG emissions. The ultimate magnitude of future climate change, however, is highly dependent on the extent of future GHG emissions.

According to scenarios developed by climate scientists, substantially lower man-made GHG emissions would alter the climate forecast, limiting increases in global temperatures and the severity of impacts related to climate change (IPCC 2007a). This chapter briefly reviews GHG sources to provide a sense of how such emission reductions can be achieved. Recommended reduction targets for global emissions are discussed, to provide a context for understanding U.S. reduction targets. Guidance is also provided on how local communities can assess their own emissions and establish their own reduction targets.

GREENHOUSE GAS SOURCES

Carbon Dioxide from Burning Fossil Fuels

Carbon dioxide is the most significant GHG, accounting for an estimated 77 percent of global GHG emissions from human activities (IPCC 2007a) and 83 percent of all U.S. GHG emissions (EPA 2009b). Fossil-fuel use for energy accounts for 93 percent of U.S. CO_2 emissions; hence, energy-efficiency and renewable-energy strategies discussed in Chapter 2 are critical to GHG reduction, as well as to long-term energy sustainability. Because CO_2 is the most prevalent GHG, other GHGs are often expressed in terms of their carbon-dioxide equivalent (CO_2e) in order to provide a relative comparison of their impacts.

As oil, coal, and natural gas account for virtually all fossil-fuel use, they also account for the majority of CO_2 emissions. Oil contributes the largest share (43 percent of U.S. energy-related GHGs). Coal, while accounting for about 22 percent of U.S. energy consumption, contributes 36 percent of all U.S. energy-related CO_2. Due to its chemical composition, coal is a more carbon-dense fuel, generating over 50 percent more CO_2 per unit of energy, on average, than oil does. Carbon intensity also varies by the type and quality of coal—certain types of coal are wetter or have more impurities. Natural gas, on the other hand, produces about half as much CO_2 as coal (EPA n.d.) Hence, while use of natural gas also needs to be reduced in the long term, it can be an important transitional energy source.

Carbon Dioxide from Land-Use Change and Agricultural Practices

Land-use change is the second-largest source of global CO_2 emissions from human activities (IPCC 2007a). Deforestation and land clearing greatly reduce the capacity of natural vegetation and soils to store carbon. The subsequent burning and decomposition of wood and organic matter turn what was a carbon sink that had helped offset other carbon emissions into a significant and widespread carbon emissions source. This does not mean that it is always undesirable to remove wood from forests. Good forest management and hazardous-fuels reduction, which include removal of trees and underbrush, are important tools to keep forests healthy and reduce the risk of wildfires. Sustainably managed forests provide important products and economic benefits. It can be important to make forest retention economically beneficial for a landowner. Still, careful attention needs to be paid to the best use of land from a carbon-emissions perspective, taking into account the multiple benefits derived from various land uses.

On the bright side, reforestation and land management in some parts of the world, including North America, are providing an important carbon sink and storage reservoir. Annual net U.S. emissions would be nearly 1 billion tons, or about 15 percent greater, if not for sequestration by forests and grasslands (EPA 2009b).

Agricultural practices also either store or release carbon in farmland soils, adding to or subtracting from net emissions. Hence, management of agricultural land can support an emission-reduction strategy (IPCC 2007a).

Figure 4.1. The burning of fossil fuels generates emissions.

Figure 4.2. A woodland fire in Wyoming

See Chapter 11 for more information about the importance of agriculture and forest management as a carbon reduction strategy.

Carbon Dioxide from Cement Production

The process of producing and curing cement emits significant amounts of CO_2. Cement production is a particularly significant source in rapidly developing countries, more so than in the United States, where urban development is already at an advanced stage. Cumulative concrete-related emissions account for an estimated 3.4 percent of global CO_2 emissions (Hanle et al. n.d.).

Methane

Methane is the second most significant greenhouse gas, contributing the equivalent of 9 percent of U.S. GHG emissions (EPA 2009b). Annual emissions and total concentrations of methane are much smaller than carbon dioxide, but methane is a more powerful greenhouse gas—with 34 times the 100-year global warming potential of CO_2. In addition, methane is relatively short-lived in the atmosphere, with an approximate residence time of 10 to 20 years, which means its near-term impacts are even greater relative to CO_2. (CO_2 is a relatively weak but abundant and long-lived GHG, lasting 100 to 200 years in the atmosphere, on balance [IPCC 2007b].)

Human-created methane is emitted by three primary sources: farm animal operations, certain crop production (primarily rice paddies), and landfills. Farm animals, cattle in particular, emit methane through their digestive process. Methane is also released from the decomposition of their manure. Similarly, the organic matter in municipal solid waste also creates methane when it decomposes, typically in a landfill. Rice production produces methane via the decomposition of organic matter in a low-oxygen environment.

Natural wetlands also tend to emit methane for similar reasons, but these are not man-made emissions. There are also large deposits of CO_2 and methane in northern permafrost. Researchers now estimate the amount of frozen carbon at 1.5 trillion tons (Global Carbon Project 2009). Warming temperatures would cause the permafrost to melt faster, releasing more CO_2 and methane, which in turn would cause additional warming—a potentially dangerous feedback loop.

Nitrous Oxide

Nitrous oxide is the third most significant GHG emitted by human activity. It is both found naturally and created artificially. Large-scale emissions come primarily from two sources: application of nitrogen-based fertilizers and combustion of fossil fuels. Strategies to reduce CO_2 emissions from fossil-fuel use also address nitrous oxide, so reduction strategies specific to nitrous oxide are focused on agricultural sources—primarily through alternative agricultural soil and fertilizer management (EPA 2009c).

Hydrofluorocarbons and Other High Global Warming–Potential Gases

Synthetic gases such as hydrofluorocarbons (HFCs) and others are emitted into the atmosphere in very small quantities but have global warming potentials that range from a few hundred to several thousand times greater than CO_2. HFCs are used in refrigeration and air-conditioning (in place of chlorofluorocarbons, earlier refrigerants that depleted the ozone layer). Several other gases are the by-products of different industrial processes. The combined emissions from these gases account for approximately 3 percent of U.S. GHGs (EPA 2009b).

Natural Greenhouse Gas Sources

Greenhouse gases occur naturally. The planet would be approximately 57°F colder on average if not for the natural levels of CO_2, methane, nitrous oxide, and other GHGs in the atmosphere (IPCC 2007b). The most important GHG, technically speaking, is water vapor, which is very abundant in the atmosphere and has a tremendous capacity to trap heat. The interaction between water vapor and other GHGs is complex and is the subject of ongoing research. Put simply, CO_2 and other GHGs interact with and influence water vapor to amplify the greenhouse effect. Ozone in the upper layers of the atmosphere, in addition to filtering ultraviolet rays from the sun, also traps heat radiating from the earth. Ground-level ozone emitted from fossil-fuel consumption adds minimally to the effect of this high-altitude ozone.

GLOBAL GHG REDUCTION TARGETS

Total global GHG emissions currently stand at approximately 40 billion tons of CO_2e per year. Scientists estimate that approximately 9 to 10 billion tons of CO_2 are absorbed by the oceans and land-based ecosystems annually (IPCC 2007c). Emissions above that threshold increase the total concentration of GHGs in the atmosphere, which are typically expressed in parts per million (ppm). GHGs make up a small fraction of the gases in the atmosphere but have a large effect on how much heat is trapped by it. CO_2 concentrations are currently about 385 ppm, rising about 1 to 2 ppm per year. Methane concentrations are approximately 1.745 ppm. Total GHG concentrations were estimated at the equivalent of 433 ppm of CO_2 in 2006 (Blasing 2009).

Scenarios analyzed in the most recent report of the Intergovernmental Panel on Climate Change indicate that stabilizing CO_2 concentrations at approximately 400 ppm and total GHG concentrations between 445 and 490 ppm CO_2e would limit increases in global temperatures from preindustrial times to the range of 3.6 to 4.0°F. Global temperatures have already risen almost 1.5°F. More recent studies (Hansen et al. 2008) have suggested that CO_2 actually needs to be stabilized closer to 350 ppm, accompanied by reductions in other GHGs.

Climate models indicate that to stabilize GHGs at these levels global GHG emissions would need to be reduced 50 to 85 percent below 2000 levels by 2050 (IPCC 2007b). The IPCC report suggests that global emissions must peak by about 2015 in order to be on a successful reduction path. Interim reduction goals may be needed, given that later emission reductions may be more difficult to achieve than earlier reductions.

U.S. GHG REDUCTION TARGETS

Gross U.S. GHG emissions currently total approximately 7.1 billion metric tons (EPA 2009b). Carbon sinks provided by forests and farmland equal slightly less than 1 billion tons for a net total of about 6.2 billion tons CO_2e (EPA 2009b). These "direct" emissions currently account for about 19 percent of global man-made greenhouse emissions. "Direct" emissions are specified because in a global economy where goods and energy flow readily among countries and the United States is a net importer, it is difficult to attribute emissions precisely to a given country. Hence, all countries may ultimately have to address this issue. For the United States, an 85 percent reduction in all GHG sources would mean reducing net emissions from 6.2 billion tons to approximately 1 billion tons of CO_2e. Of the 7.1 billion tons of gross emissions, about 6 billion tons are energy related.

There are also end uses associated with each of these fuels. For local energy and GHG-reduction strategies, it is helpful to focus on end uses, as that is where the greatest efficiency and renewable energy opportunities ultimately lie. Figure 4.3 shows the breakdown of U.S. GHG emissions associated with different end uses (EIA 2009c).

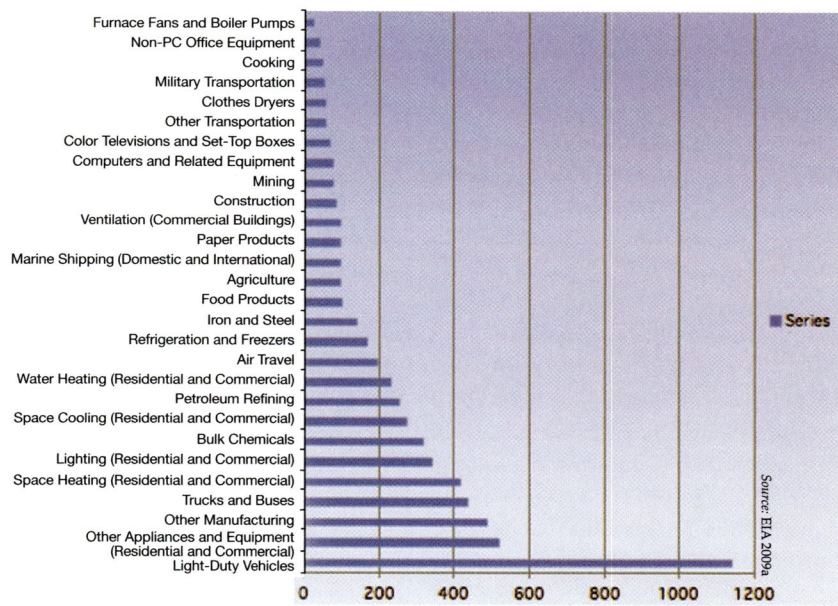

Figure 4.3. U.S. energy-related CO_2 emissions by end use (2007), in millions of metric tons

LOCAL GHG EMISSION REDUCTIONS

Local GHG reduction targets will generally need to parallel national targets. Local GHG reduction strategies will need to find potential reductions among different end uses, either by reducing energy use through efficiency or by shifting to renewable or lower-carbon energy sources. As noted, however, local energy use and GHG emission profiles can vary widely because of regional and state differences, as well as different actions taken at the community level. For example, communities may get a greater or lesser share of their electricity from coal-fueled power plants versus less carbon-intensive sources. Energy demand for heating and cooling can vary depending on local climate as well as local building practices and history. The energy intensity of local businesses can differ over a wide range depending on the type of industries and different practices by companies within the same industry.

A Local Greenhouse Gas Inventory

An actual estimate of energy use and GHG emissions in each local community—whether an individual town or a larger region—is an important part of developing a local GHG-reduction strategy. Comprehensive protocols for GHG inventories have been developed for individual businesses, facilities, and municipal operations, as well as individual U.S. states and countries. Protocols for community-wide GHG inventories are also being developed. Other entities have created tools for local communities interested in doing a complete inventory.

There are a number of practical challenges and issues that communities need to consider and address in measuring a local GHG footprint and tracking progress on reducing emissions over time. Defining the boundaries of a local GHG footprint—that is, what will be included or excluded in an accounting—is an important first step. This can be difficult, since some emissions are direct—created by an activity occurring locally—and some are indirect, meaning they are created elsewhere but associated with or connected to local economic activity in some way. Also, transportation emissions are mobile sources that must be accounted for somehow.

Determining which emissions are a community's responsibility and which are another community's or someone else's responsibility are problems that arise frequently. For the sake of keeping track of the progress made toward an established reduction target, defining what is included and what is not can be helpful. Ultimately, however, communities will likely have opportunities to reduce both emissions that are clearly their own as well as emissions that arguably are from somewhere else. If desired, those emissions can be tracked separately as "extra credit," or all GHG-reduction activities can be totaled and tracked. This will depend, naturally, on how a community defines the scope of its emissions inventory.

Established GHG inventory protocols generally define emissions as scope 1, scope 2, or scope 3 emissions (California Climate Action Registry 2009). Scope 1 emissions are those generated within a defined area—such as those from heating-fuel use, gasoline consumption, or a local industrial process. Scope 2 emissions occur outside the local area but arise from local activities, such as those from the use of electricity that is generated elsewhere and then imported. Scope 3 activities are indirect emissions created or generated by an activity that occurs outside the local area.

Transportation is a special case, and the boundary of transportation emissions is often defined by available data. A reasonable definition, though, includes all the fuel used for commuting and other passenger trips by people residing within the area, as well as noncommuting work-related trips for employees originating within the local area. In addition, freight trips terminating within the local area for goods or raw materials are included.

Here are some examples of how different localities or organizations approached the inventory problem. New York City decided to measure all vehicle travel within the city limit because that corresponded with the data available to the analysts. Inventorying a metro area can be a special challenge because energy and GHG-related data is not always collected at the regional level. Hence, a study by the Brookings Institution that sought to compare the carbon footprints of different metro areas chose, for practical reasons, to perform a partial inventory that included electric use by residential and commercial buildings and regional passenger-transportation use (Brown et al. 2008). The study omitted the industrial sector entirely as well as freight, which was reasonable given its purpose of creating a relative comparison between areas and under defined parameters.

The important challenge for any community is to be clear and consistent about how it defines its total GHG footprint and its different subelements and what kind of data it uses as inputs. This will ensure consistent tracking compatibility for trend measurement. For example, if records from the local gas or electric company are used as a primary source of information, it will be important to continue to use that source of information and method of estimation in future GHG assessments. It may be helpful to update or introduce software to manage data more easily or to start tracking new information, but it is important to be able to keep track of the old data in order to chart a valid trend. For example, if an analysis starts by using electric bills as input data but then later includes customer surveys, the two bodies of data may not be compatible or comparable, and their value in measuring trends will be lost or reduced.

The purpose of local GHG inventories is to identify the most promising reduction opportunities and set priorities. Precision is not as important as comparing relative cost-effectiveness and potential size of different reduction options. Some type of inventory is helpful to start, but some communities get preoccupied in the assessment phase and do not proceed to the action phase. Action is really the most important part, to which we turn next.

CHAPTER 5

Strategic Points of Intervention

- **Long-Range Community Visioning and Goal Setting**
- **Plan Making**
- **Standards, Policies, and Incentives**
- **Development Work**
- **Public Investment**

Communities seeking to respond to energy and climate challenges are often faced with two daunting questions: What do we do and where do we start? The enormity and complexity of the issues may cause some communities to put off addressing these challenges, while other places may choose to create a green team or initiate a climate action plan process yet otherwise conduct business as usual.

Planners have many opportunities in the planning process and in their day-to-day work to make a difference on energy and climate issues. Whether reviewing a project application or updating a comprehensive plan, planners should consider how proposed plans and projects can reduce energy use and greenhouse gas (GHG) emissions and prepare to adapt to a changing climate.

This chapter describes what planners do and where in the planning process opportunities exist to effect change in response to a particular issue. These strategic points of intervention should also be communicated to those who may not be as familiar with the planning and community development process.

LONG-RANGE COMMUNITY VISIONING AND GOAL SETTING

Planners often conduct visioning exercises that produce long-term goals and objectives that community leaders look to when considering policies and actions. Community visioning can be the first step in developing a comprehensive, neighborhood, or downtown plan. Whether part of a planning process or on its own, visioning is an important first chance to identify new opportunities and priorities related to energy and climate.

Here are some ideas for how planners can integrate energy and climate issues into the visioning process:

- *Survey citizen attitudes.* Gauge the level of awareness and importance of energy and climate change issues to community members. In a community survey, for instance, ask questions such as, How concerned are you about greenhouse gas emissions in the community?

- *Hold community workshops.* Consider how energy and climate issues can be addressed through interactive forums. For example, in community workshops, create an exercise to gauge the level of support for renewable energy options such as solar panels and wind turbines.

Fig. 5.1. A community workshop

- *Connect community goals.* Determine how energy and climate objectives are connected to other community goals and values. Review a list of these goals and values or a draft vision statement. For example, is a walkable, bikeable community part of the vision? Work toward such goals can also help reduce GHG emissions from transportation. Discuss these connections with the community.

- *Run alternative scenarios.* Show carbon-footprint consequences of alternative land-use policies using GIS-based tools.

- *Task force appointment.* Consider including an energy/climate change task force or committee in visioning and goal-setting work.

Communities may also use a visioning process to discuss setting new energy and climate goals or to brainstorm ideas for meeting new goals and targets. For example, if a community or state has set a target for GHG emission reduction or signed on to an agreement that does, the visioning process can be a good venue for discussing ideas about how to start meeting those goals or interim targets. If a community does not yet have a GHG emission-reduction goal, the visioning and goal-setting process can be used to discuss setting one. Planners can also provide baseline scientific and technical information on climate change to help inform a community. What target a community decides to set or accept might ultimately be determined by its elected officials, but buy-in from the community can be important, especially when trying to meet those goals. Citizen interest is often one of the largest motivators for government action.

While there are communities that are currently not interested in or resistant to exploring GHG emissions-reduction strategies, these views may change.

Moreover, many of the things that a community does accept or promote can play positive roles in reducing emissions and mitigating climate change. Identifying and communicating the extent of these existing activities in reducing emissions may be one way to help overcome community indifference.

PLAN MAKING

Planning departments prepare plans of all kinds. They recommend actions involving infrastructure and facilities, land-use patterns, open space, transportation options, housing choice and affordability, and much more. Examining comprehensive plans and other planning documents to see if energy and climate change issues are addressed and integrated is an important step.

Assessment and Analysis

An initial step to almost any planning process is a baseline assessment and analysis of existing conditions and trends. Establishing the baseline for energy use and GHG emissions is critical to be able to track and measure progress toward emission-reduction goals. A good baseline measure of a community's energy use will take account not just of the amount used but also the mix of renewable and nonrenewable sources. Similarly, a GHG inventory (see Chapter 4) will provide an assessment of the quantity and source of emissions in your community.

An assessment of potential new risks and impacts associated with climate change will also be helpful in planning for these impacts and considering adaptation measures. Increasingly, information on expected climate-change impacts is being provided at a regional scale (as discussed in Chapter 3). Federal and state agencies, universities, and other organizations in a given region may have already conducted detailed assessments of anticipated impacts there. State climatologists and regional climate centers are also good sources of locally applicable climate-change information.

These assessments can be summarized and included in an overall plan for easy reference and connection to goals and policies that respond to them. This baseline information can also be used to set goals for emission reduction and renewable energy use, if they are not already in place. However, the task of comprehensive assessment should not hinder progress in other areas. Information with which to make energy and climate assessments may vary in quality and availability, but data precision is generally less important, at first, than establishing relative priorities. Going through the process, however, can help identify information gaps and data-collection needs.

Comprehensive Plans

The comprehensive plan is a guiding document for the future of an entire community. It establishes goals and priorities and lays out action steps for meeting those goals. Accordingly, the importance of addressing energy and climate in the comprehensive plan should not be overlooked.

Planners should consider including an energy and climate change element in the comprehensive plan, integrating these issues within other elements, or both. Devoting an element to these issues may provide focus and allow communities to more easily amend an existing comprehensive plan. However, planners should also consider how energy and climate issues relate to other issues and elements in the comprehensive plan, such as land use and transportation. If a community considers these issues especially important, it may be advisable to address virtually all comprehensive plan issues and elements under the umbrella of an emissions reduction or sustainability vision. (See the case study on Marin County, p. 50.)

MARIN COUNTY, CALIFORNIA, COUNTYWIDE PLAN

Population: 248,794

Marin County has a decades-long history of incorporating environmental concerns into its land use decisions. It adopted its first formal countywide plan in 1973. Residents are proud of the county's natural resources and are generally receptive to issues of conservation, environmental protection, and climate change responses. It is thus no surprise that in its most recent plan update, the county incorporated climate change mitigation and adaptation efforts into land use, energy conservation, green building, transportation, waste disposal, and most of the other topics included in the plan.

County officials became interested in addressing sustainability and climate change issues in the comprehensive plan because there was not much leadership on climate change coming from state or federal sources. (This was six years before the passage of AB 32, California's Global Warming Solutions Act of 2006, which requires the state to reduce GHG emissions to 1990 levels by 2020.) The general plan, as dictated by California state law, guides the physical development of the county and has a major impact on both urban form and management of resources. In Marin's case, it also addresses such social equity and cultural issues as public health, environmental justice, child care, the economy, and arts and culture. The updated countywide plan was developed by the Marin County Community Development Agency and adopted by the Marin County Board of Supervisors on November 6, 2007.

County staff began the general plan update process in 2000. They chose "planning sustainable communities" as the plan's theme. Accordingly, sustainability principles are incorporated into the entire plan and not just added as a stand-alone policy.

The countywide plan involved extensive education and background work, as well as coordination among county residents, agencies, departments, and administrators. Early on, the Community Development Agency teamed with the county administrator and Board of Supervisors to modify the county's mission statement; the new statement links excellence in public service with the support of sustainable communities in order to preserve Marin's unique environmental heritage. The County Strategic Plan was also updated to include the goal of sustainable communities. In 2002, the board signed on to the Cities for Climate Protection program developed by ICLEI–Local Governments for Sustainability, conducted a GHG emissions analysis, and set emissions-reduction goals.

As part of the process, a sustainability working group of local residents was convened to help prepare guiding principles. The resulting 12 principles form the basis for 11 countywide goals that are the base layer for all policies, programs, and implementation measures of the plan. Both the principles and the goals contain several items that explicitly address the need to change land use, development, and transportation patterns in order to lower GHG emissions and slow the rate of climate change. Other principles and

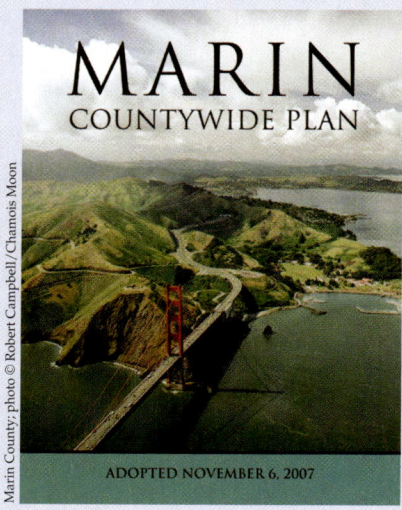

Fig. 5.2. The cover of the Marin countywide plan

goals address aspects of sustainable communities that are related to climate change and energy such as healthy local food, protected water resources, intact ecosystems, and well-designed, affordable housing built near transit nodes.

The countywide plan is organized into three elements: natural systems and agriculture, the built environment, and socioeconomics. Many of the climate change policies and programs are discussed in the atmosphere and climate chapter in the natural systems and agriculture element, as well as in the energy and green building chapter in the built environment element. The policies aim to reduce GHG emissions, monitor climate change, and adapt to climate change's effects. Some of the ways the plan mandates reaching these goals include increasing the use of renewable energy; developing green building and energy-efficiency programs; changing commuting and driving patterns; and reducing methane emissions from solid waste disposal. Adaptation measures include revision and implementation of the floodplain ordinance; increased research on sea-level rises in the county; and establishment of a climate change planning process.

To bring cohesion and consistency to the document, a format of four questions is used to address each goal in the plan:

▶ What are the desired outcomes?

▶ Why is this important?

▶ How will results be achieved?

▶ How will success be measured?

The last two questions especially are used to formulate indicators, benchmarks, and targets to help measure and evaluate progress. Each section of the plan is also followed by a program implementation table, which is designed to be used as the basis for applicable performance plans and work programs that are part of the county's annual budget-setting process. The table summarizes responsibilities, potential funding priorities, and estimated time frames for proposed implementation programs, with the time frames dependent upon the availability of adequate funding and staff resources.

(continued on page 51)

In some states where guidance is provided on preparing local comprehensive plans, some agencies and organizations are starting to suggest how to address these issues. The California Air Pollution Control Officers Association, for example, released a white paper in June 2009 that contains model comprehensive plan policies for reducing GHG emissions.

Here is some general guidance on addressing energy and climate issues in comprehensive plans:

- Provide a narrative description of the issue and the rationale for addressing energy and climate in the plan

- Summarize existing conditions and any baseline assessments

- Establish broad goals related to energy and climate

- Develop policies that support these goals and connect to land use, transportation, infrastructure, housing, economic development, and other important aspects of the plan

- Create implementation or action steps to achieve energy and climate-related goals, identifying who will lead the implementation, what the timeline will be, and any known resources or funding sources that could help with the implementation.

Planners should also review existing and proposed aspects of the comprehensive plan to determine if there are conflicting goals or policies. For example, if reducing vehicle-miles traveled is a goal, does the land use element support it by promoting a mix of land uses and compact development patterns? Does the transportation element support modes of transportation with fewer GHG emissions, such as transit and bicycling? Are bicycle lanes included in the plan? Looking for possible connections or conflicts is critical to creating a coherent strategy for a new energy and climate future. Chapters 6 through 11 include guidance on opportunities for reducing energy use and GHG emissions in important issue areas.

Area Plans

In addition to a comprehensive plan, many communities also have more specific area plans. These may include neighborhood plans, downtown plans, redevelopment district plans, and corridor plans. Opportunities to address energy and climate concerns also exist in area plans. Some general guidance for incorporating these issues into area plans includes:

- Describe the issue, as in the comprehensive plan, but link it to the specific planning area to the extent possible

- Reference related goals and policies in the comprehensive plan

- Create more specific yet consistent recommendations for energy and climate policies in the particular area.

Functional Plans

Functional plans focus on a single issue or set of issues rather than a geographic area. They may apply to an entire community or part of a community. Transportation plans, including plans for streets and sidewalks, transit, bicycle facilities, and pedestrian circulation, as well as airport master plans, are examples. Others include open space plans, parks and recreation plans, sewer and water plans, school plans, and

---

(continued from page 50)

One of the interesting aspects of the Marin plan is that many of the policies and programs in it are partially funded because they are part of the ongoing operations of the county. Thus, incorporating new aspects to make them more "green" or sustainable is not a matter of starting from scratch but rather adjusting departmental operations or other programs to reflect the sustainability goals. However, those changes do not come free. Alex Hinds, former director of the Marin County Community Development Agency, says, "Integrating sustainability and climate change issues into the County's business practice requires paying someone to do the work."

Hinds explains that the county was able to use some of the money set aside for updating the general plan to implement the energy and climate change programs, and it relied upon existing staff and both paid and unpaid interns to stretch dollars farther. Marin County was able to use some funds set aside for preparation of the general plan by concurrently preparing and implementing green strategies and the plan. The county also added modest increases to land use and building permit fees to provide for the ongoing task of reviewing proposed projects to ensure that green objectives such as construction waste recycling are accomplished.

Other sources of implementation funding come from a variety of grants, especially those provided by local utilities and state agencies concerned with waste management and energy, and some general fund and occasional foundation monies. The recently enacted federal Energy Efficiency and Conservation Block Grant Program may also be a source of additional funding.

Marin County has many notable environmental assets: the rocky Pacific coast headlands at Point Reyes National Seashore; the towering redwoods of Muir Woods State Park; hundreds of endemic or endangered species; and a thriving local agricultural industry, to name a few. The countywide plan serves to help preserve those assets and protect the county from some of the effects of climate change, while curbing the county's greenhouse gas footprint. ◀

economic development plans. Some functional plans are prepared by a municipality or a public or private special-purpose entity, such as a utility, an authority, or a school district.

All such plans should consider how they might affect energy use and GHG emissions. Does the transportation plan support increased use of more energy-efficient modes of travel, such as transit and bicycling? Does the open space plan include plans for conserving or increasing natural areas, which are important for carbon sequestration? Does the economic development plan foster green industry? Asking these and similar questions can help identify where efforts can be improved.

Climate Action Plans

Climate action plans (CAPs) provide policy direction for reducing GHG emissions. They generally include baseline information on GHG emissions, targets for reducing these emissions, and strategies for achieving GHG-reduction goals. They may also include timeframes, milestones, and tracking and accountability measures. Some also include adaptation strategies to respond to the projected impacts of climate change. (See the Berkeley case study.)

CAPs can be useful for focusing on GHG emissions and strategies to respond to climate change. When prepared prior to a comprehensive plan update, they can provide much of the baseline information needed to incorporate appropriate GHG policies into the update (CAPCOA 2009). But while these plans can be useful tools, they alone may not be enough to fully address energy and climate issues. Comprehensive plans and zoning ordinances may need to be updated to ensure consistency with a CAP's goals. Other mechanisms for implementing a CAP will likely need to be put in place.

While CAPs have limitations by nature, they also vary widely in content and quality. Many lack adequate strategies and measures, few address adaptation, and implementation is problematic (Wheeler 2008). They also vary greatly in content, with many focusing on municipal strategies such as greening vehicle fleets and public buildings without addressing important areas in land use and transportation where planners can help make a difference. This may be due in part to the fact that some of these plans have been prepared outside of planning departments, sometimes with little input from planners. When possible, planners should take a leading or active role in the preparation of CAPs to ensure a more comprehensive approach and to fully integrate recommended actions into the strategic points of intervention.

CAPs are becoming more widespread and can be useful tools, but they are not panaceas for all the climate-related challenges that communities face. Without important changes to the comprehensive plan, the zoning code, and other implementation mechanisms, CAPs run the risk of being just more plans on the shelf, collecting dust or taking up file space.

Energy Plans

Energy plans typically provide an overview of energy use and sources in a community, as well as strategies for ensuring energy security in the future. Like CAPs, energy plans can be helpful for focusing on and addressing the issue. While energy plans have been in place in some communities for many years, they are becoming more common as the importance of addressing energy becomes more apparent and as access to funding opportunities for preparing such plans increases. As with CAPs, communities should ensure consistency among plans and take appropriate steps to implement an energy plan's recommendations.

BERKELEY, CALIFORNIA, CLIMATE ACTION PLAN

Population: 101,555

In November 2006, Berkeley voters issued a call to action on climate change: 81 percent of them endorsed ballot Measure G, which mandates that the entire community's GHG emissions be reduced to 80 percent below 2000 levels by the year 2050. Two aspects of this vote stand out: the GHG reductions apply to everyone in Berkeley, not just city operations and facilities; and GHG reduction is something the city takes very seriously. Measure G set a local climate-protection campaign in motion. The Berkeley climate action plan (CAP) is designed to serve as the city's guide in making the campaign a success.

The CAP was adopted by the Berkeley City Council on June 2, 2009. It took time to draft and finish the plan, but the result is a comprehensive document developed in a collaborative manner. The council allocated two years of funding for research on climate-protection strategies and to conduct a robust community-input process. Development of the plan was a cross-departmental effort coordinated by the city's Office of Energy and Sustainable Development (OESD). OESD relied on the expertise of staff from the Department of Public Works, which includes the Transportation Division and the Solid Waste Management Division; the Department of Planning and Development; the City Manager's Office, which includes the Office of Economic Development and Neighborhood Services staff; and the Department of Health and Human Services, among others.

The public process was designed to maximize the opportunities community members had to contribute ideas, learn more about climate change issues, and get involved in existing sustainability efforts. Some of the opportunities for public participation prior to the release of the first draft in 2008 were:

- *Climate Action Kick-Off:* Held in May 2007 and attended by more than 170 community members.
- *Commission-hosted Climate Action Workshops:* Seven city commissions hosted public workshops for the purpose of providing a forum for participation in plan development.
- *Community events and meetings:* City staff and volunteers participated in many community events. More than 1,500 people stopped by a Berkeley Climate Action booth or attended a community event with a climate action component.

Local experts in the fields of climate science, energy, transportation, and public engagement were appointed as informal advisors. The city also took advantage of the University of California's flagship campus by welcoming contributions of research, labor, and guidance by UC faculty, staff members, and student leaders.

The CAP focuses on three main topics: sustainable transportation and land use; building energy-efficiency and usage strategies; and waste reduction and recycling. It also contains chapters on the Berkeley GHG inventory; community outreach and education; adaptation to climate change; and overall implementation of the plan.

One of the most important (perhaps *the* most important) aspect of any climate change plan is how well the plan can or will be implemented. The Berkeley CAP does a good job of detailing this. Each chapter contains a table (all of which are also in the CAP's Appendix A) that lists each action or goal and indicates the department or governmental unit responsible for implementation, the entity that provided funding, and whether the goal is to be implemented in the short, mid-, or long term.

Fig. 5.3. Photovoltaic solar panels on university senior housing, Berkeley, California

Timothy Burroughs, the climate action coordinator for the City of Berkeley, says, "In essence, the Berkeley CAP serves as a point of departure for securing funds.... [T]he CAP articulates the City's strategic thinking related to how to effectively reduce GHGs and achieve other co-benefits at the local level. With much of this strategic thinking done, though always evolving, the City can target sources of funding to support implementation. The CAP reflects the City's implementation priorities as well as the analysis behind those priorities."

For Berkeley, local climate action is a top priority. The city is dedicating about $1.3 million of its general fund to implementation of the CAP. This includes funds for outreach and education, monitoring and reporting CAP implementation, policy development, and more. Burroughs also states that staff is working to secure funding for CAP implementation through grant writing and applying for federal stimulus funds, such as the Energy Efficiency and Conservation Block Grants (EECBG), of which Berkeley was allocated about $1 million. Securing that funding requires the city to submit a proposal and an energy-efficiency and conservation strategy. Berkeley's goal is to utilize EECBG funds to provide incentives (in the form of rebates) to residents and businesses for the purpose of energy audits and efficiency

(continued on page 54)

(continued from page 53)

improvements. Through a separate U.S. Department of Energy funding opportunity, the city is also applying for funds to put solar energy systems on City Hall and the city's planned new animal shelter.

The city is also seeking federal funds through the U.S. Environmental Protection Agency to support monitoring and reporting on implementation of the CAP. Specifically, staff intends to establish, monitor, and report on progress indicators that will help the city and its partners to effectively communicate CAP implementation progress to the public. The EPA grant would enable staff to build on existing monitoring and reporting efforts. The OESD has already identified an initial list of progress indicators related to transportation, land use, building energy use, waste diversion, and outreach and education, with indicators for other categories planned. Though reliable data is only sometimes available for these indicators, the OESD wants to find ways to make them accessible to the community, funding agencies, and others. To this end, the city is using a web application called See-It, developed by Visible Strategies, to communicate the goals of the CAP as well as progress toward achieving its goals. The city sees sustained and transparent reporting on implementation of the CAP, as well as on the effects of that implementation, as fundamental to the success of the local climate-action effort.

Berkeley is also moving forward with existing projects and programs that are consistent with the CAP. According to Burroughs, Planning Department staff is working with community stakeholders to finalize an updated Downtown Area Plan and Southside Plan, as well as to provide zoning flexibility within the West Berkeley Plan. A fundamental component of each of these efforts is to prioritize a mix of uses and access to transit and other amenities. The city is also focused on expanding its bicycle infrastructure. In addition to seeking grants and continuing work on program development and implementation that is consistent with the goals of the CAP, the city is interested in identifying sustained sources of revenue that can support transportation demand management (TDM) efforts and others. These include implementing parking strategies that create disincentives for driving and, where possible, provide revenue for alternative transportation, and developing a transportation services fee for new development to provide revenue for TDM efforts.

The City of Berkeley and its residents have taken the threat of climate change seriously and are moving forward on a number of fronts to mitigate climate change impacts and adapt to the future. The Berkeley CAP is an example of how these plans can articulate a vision and provide a solid framework for action at the same time. ◄

STANDARDS, POLICIES, AND INCENTIVES

Planners write and amend standards, policies, and incentives that have an important influence on what, where, and how things get built and what, where, and how land and buildings get preserved. When updating regulations, planners should consider how zoning codes, building codes, subdivision codes, and other regulations and ordinances address energy and climate issues and how these could work to encourage energy-efficient and climate-friendly forms of development.

There are three main steps to revising a code to address energy and climate issues (Duerksen 2008):

1. Remove barriers: existing codes may inhibit reducing GHG emissions and using renewable energy, often unintentionally. The planning office can conduct a barrier audit to determine what regulatory obstacles exist. Some codes may prohibit wind turbines, for example, due to height restrictions. Other ordinances may prohibit mixed uses, accessory dwellings, and higher residential densities.

2. Enact standards: setting appropriate standards to guide desired development is important. For example, standards can mandate the preservation of trees (which aid in carbon sequestration) and require new subdivisions and mixed-use development to include bike lanes and sidewalks.

3. Create incentives: desired types of development can be fostered through incentives. For example, development that includes green roofs (which can help reduce the urban heat island effect while reducing heating and cooling loads in buildings) could receive a density bonus. Both the added density and green roofs could help a community meet its energy and climate goals.

Zoning Code

Perhaps the most important regulatory tool for development in a community, the zoning code typically establishes permitted uses in various locations and provides standards for intensity of use, such as lot size, floor area ratio, setbacks, building heights, and permitted accessory structures. Some recent forms of zoning codes, such as form-based codes, emphasize building form over use. In reviewing and revising a zoning code to address energy and climate issues, some things to look for include:

- Minimizing impervious surfaces to reduce heat island effects and runoff
- Mixing land uses to shorten and reduce vehicle trips
- Increasing development densities, especially around transit
- Accommodating solar orientation
- Allowing a variety of housing types to reduce work trips

- Reducing parking requirements through shared parking
- Requiring landscaping, mature tree preservation, and open space.

Subdivision Regulations

Many communities also have subdivision regulations that include standards for the design and layout of lots, streets, and other public improvements. Again, these regulations are an important tool in promoting energy-efficient and climate-friendly design and development. Regulations should promote street connectivity, require sidewalks and bike lanes or paths, and protect environmentally sensitive areas. Other pertinent development standards include block standards, right-of-way width, roadway design, and stormwater management and open space standards.

Planned Unit Development Regulations

Planned unit development (PUD) regulations often allow more flexibility than zoning and subdivision regulations do. They are commonly used for approving master planned communities and mixed use developments, though a PUD can take different forms. Some PUD regulations include a number of development standards, while others minimize prescriptive standards and use a discretionary review process to allow more flexibility in design. A PUD can often allow for mixing of land uses, densities, and housing types; more compact development; and preservation of open space that can help meet energy and climate goals. Planners should review PUD regulations and encourage PUDs that are responsive to energy and climate issues.

Incentives

There are a number of incentives available from federal and state sources, as well as utilities, that promote energy efficiency and reduced carbon emissions. The Database of State Incentives for Renewable Energy (DSIRE), for example, provides links to information on many incentive programs that promote renewable energy and energy efficiency (www.dsireusa.org). Planners should consider creating a fact sheet with information on available incentives for developers and residents, as these incentives can be helpful in meeting energy and climate goals.

Local governments also can create their own incentives. These may include expedited plan review for projects that meet or exceed established energy and climate objectives; a waiver of permit fees, rebates, or other financial incentives to developers whose projects meet predetermined standards; and provision of technical assistance to help developers meet new goals and standards. Some local governments help residents and businesses to invest in renewable-energy improvements such as solar panels through a financing mechanism that works like a sidewalk assessment. Cities that own their own utility also can offer incentives such as rebates for purchasing energy-efficient appliances and encouraging the purchase or even the production of green power. (See the Pasadena case study, p. 56.)

DEVELOPMENT WORK

Planners play an important role in development in their community. They review project applications for consistency with applicable plans and regulations and may be involved in public-private partnerships to develop new projects. While goals and standards for energy and climate should be addressed in plans and regulations, making sure these goals and standards are met or exceeded in the development process is important.

PASADENA'S GREEN BUILDING PROGRAM

Population: 143,400

The City of Pasadena has a strong record of environmental stewardship. In 2006, for example, the city released its Green City Action Plan, which included goals to conserve energy and water, reduce waste, address global warming, protect natural habitats, improve transportation options, and reduce risks to human health.

One of Pasadena's earliest and perhaps most innovative actions was the adoption of a green building practices ordinance and accompanying project review standards. The efforts of decision makers, planners, and other city staff to encourage environmentally responsible building through incentives and a strong project development and review program has shown the city's dedication to sustainability.

The ordinance, adopted in 2005, establishes the U.S. Green Building Council's Leadership in Energy and Environmental Design (LEED) rating system as the standard for measuring and evaluating the environmental soundness of applicable buildings in Pasadena. The ordinance is part of a larger green building program that provides incentives and includes an outreach and education component.

Projects subject to the ordinance are required to comply with several mandatory building requirements that promote energy efficiency and GHG reductions, including: construction-activity pollution prevention; a 20 percent reduction in water use; minimum energy performance; and construction-waste management that diverts 50 percent of the stream from disposal.

To ensure that buildings comply with the energy and conservation goals of the program, the city requires project applicants to go through a rigorous development review process. Central to the process are checklists tailored to each building category covered by the ordinance. The requirements are designed to align closely with the LEED rating system and provide developers a clear guide on how to attain LEED credits while also complying with other Pasadena and State of California building requirements. The city planning department offers the project design team complimentary LEED Accredited Professional support services throughout the project development and review process.

THE ORDINANCE

Pasadena's ordinance requires that all city buildings of at least 5,000 square feet of new gross floor area and renovations of 15,000 square feet or more attain LEED Silver certification at a minimum. In order to achieve a LEED rating, buildings must meet credits in five main categories: sustainable sites; water efficiency; energy and atmosphere; materials and resources; and indoor environmental air quality. Achieving credits in a sixth category—innovation in design—is optional.

Though only municipal projects are required to attain official recognition, the city strongly encourages LEED certification for private buildings. The ordinance applies to the following private building categories and thresholds:

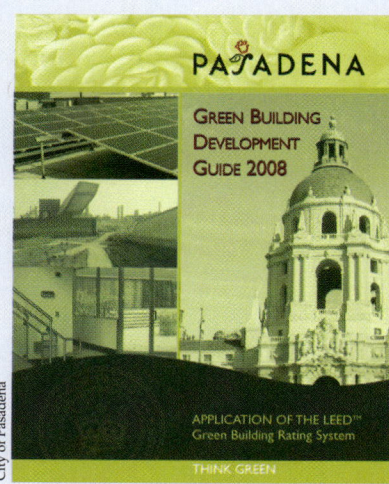

Fig. 5.4. Pasadena's Green Building Development Guide

- All nonresidential buildings of 25,000 square feet or more of new gross floor area must meet the intent of the LEED Certified level at a minimum; larger commercial or institutional buildings of 50,000 square feet or more must meet the intent of the LEED Silver level.
- Tenant improvements of 25,000 square feet or more of gross floor area and requiring a building permit as determined by the building official or designee must meet the intent of the LEED Certified level at a minimum.
- Mixed use projects and multifamily residential projects that include a residential building of four stories or more in height must meet the intent of the LEED Certified level at a minimum.

INCENTIVES

In addition to the environmental and life-cycle savings of employing green building methods, the city offers substantial financial incentives to commercial and residential developers voluntarily seeking LEED certification. The Pasadena Water and Power Department, a city-owned utility, awards rebates for high-performance building, appliance, and energy systems that contribute to reducing emissions and improving energy efficiency in Pasadena. For example, the city has established green power-purchasing and solar rebate programs for commercial and residential buildings. This table indicates select commercial and residential water conservation, energy conservation, and renewable energy rebates that are currently available.

Sector	Incentive/Rebate Program	Description	Rebate Value
Commercial			
	High Performance Building	Matching funds for energy efficiency improvements	Up to $100,000
	LEED Certification	Incentives for receiving official LEED Certification	$15,000–30,000
Residential			
	Cool Trees Rebate	Incentive to plant trees for heat island mitigation	$40–50 per tree
	Energy Star Rebate Program	Rebate for Energy Star appliances and compact florescent bulbs	Up to $150 per appliance

Project Review

In reviewing private development projects, planners assess whether standards in ordinances and regulations have been met. Thus, a checklist of energy and climate change standards or goals for new projects can be helpful. If the goals are not mandatory, an expedited plan review and permit-processing track for projects that meet or exceed those goals can be an effective tool. Planners should also review project plans to consider ways they may be redesigned to be more energy efficient, such as orienting buildings to maximize passive solar use. Project review should also be considered an opportunity to examine whether existing zoning and other development regulations require modification to support energy-efficient design.

Public-Private Partnerships

Planners may serve as leading team members on public-private partnerships, and there they have an important role to play in creating developments that are energy efficient and reduce GHG emissions. Planners might find ways to reduce the extent of impervious surfaces; facilitate placement of buildings close to transit stops, bike paths, and sidewalks; preserve trees; and encourage the use of renewable energy on-site, among other things. (See Chapter 10.)

PUBLIC INVESTMENT

When towns, cities, and counties undertake major investments in infrastructure and community facilities, planners should be involved. These public investment decisions can substantially affect the design and location of transit, streets, sidewalks, bikeways, schools, sewer and water facilities, and other public infrastructure and facilities, not to mention the impacts on energy use, GHG emissions, and climate adaptation.

Planners should encourage investments that

- Reduce overall energy use: for example, by making buildings energy efficient

- Facilitate transitions to renewable energy: for instance, by investing in solar panels or community wind

- Encourage compact settlement patterns: by determining the vehicle trip implications of where future sewer and water facilities, schools, libraries, and trip-intensive uses are located

- Help reduce GHG emissions: for instance, by providing bike paths and sidewalks

- Recognize the need to adapt to a changing climate: for example, by locating public facilities away from floodplains.

In addition, cities can lead by example, showing that energy and GHG-reduction goals can be met, such as through energy-efficient retrofits of public buildings. Private developers may then be more likely to follow and incorporate these goals in their own projects.

Capital Improvement Programs

Planners should take active and leading roles in a community's capital improvement program to ensure that public investments—including infrastructure, public buildings, and facilities—promote energy efficiency and reduce GHG emissions. Capital improvement programs typically lay out public improvements and associated costs over the next five years. The guidelines above apply in this area as well.

EDUCATION AND OUTREACH

The importance of education and outreach in meeting energy and climate goals should not be overlooked. Community education and outreach activities often happen within each of the points of intervention highlighted in this chapter. Communities also hold separate education and outreach programs for important issues. In general, planners should consider ways to engage the public in discussing energy and climate change and provide educational forums for citizens to learn how to make changes in their own lives to improve energy efficiency and reduce carbon emissions. (See the Greensburg case study, p. 58.)

In addition, planners should consider how to reach out to other agencies and stakeholders that influence and affect energy and climate goals. These might be neighboring jurisdictions, school districts, regional transportation agencies, and local utilities. Involve them early in the process, get their input and feedback on new policies and regulations, and work with them to implement new standards.

GREENSBURG, KANSAS, SUSTAINABLE COMPREHENSIVE PLAN

Population: 1,574

On May 4, 2007, an EF-5 tornado (the most severe type) with estimated winds of 205 mph hit Greensburg. It destroyed 95 percent of the town and severely damaged the remaining 5 percent. Twelve people were killed. In planning for the rebuilding effort following the tornado, the City Council passed a resolution stating that all municipal buildings would be built to LEED Platinum standards (the highest level of certification in the U.S. Green Building Council's LEED rating system), making it the first municipality in the nation to do so. Greensburg is rebuilding "green," with the goal of becoming a model sustainable rural community. The city released its sustainable comprehensive plan in May 2008. The plan is intended to guide the city's efforts to rebuild infrastructure, facilities, homes, businesses, and institutions in ways that will ensure "a socially vibrant, economically viable, and environmentally rich future." Greensburg hopes that its development goals and its inclusive plan-making process will be a model for other rural communities across the country.

agencies. Representatives of FEMA's Long-Term Community Recovery program helped build a participatory community-planning process. The intensive 12-week process brought together the FEMA team with local, state, and federal officials, business owners, civic groups, and citizens. Hundreds of people turned out for a series of community meetings on how to rebuild Greensburg and Kiowa County. Community participation provided an invaluable source of input and feedback that was used to refine and prioritize projects. By August 15, 2007, the City of Greensburg was ready to adopt the Greensburg/Kiowa County Long-Term Community Recovery (LTCR) Plan.

One goal in the LTCR Plan was to develop a comprehensive plan based on the principles of sustainability. The community had developed a recovery vision as part of the process used to create the LTCR Plan. The vision embraces the creation of a community in which people, the economy, and the environment are all supported and connected and are successful over the long term. The idea of sustainability is seen as a corollary to rural Kansan values of thrift, stewardship, and neighborliness.

Greensburg city staff, officials, and others from outside who were assisting in the recovery knew that the endeavor to build a sustainable model rural community would require extensive coordination and alignment of hundreds of stakeholder orga-

Fig. 5.5. A community meeting under the FEMA-provided tent, summer 2007

The Greensburg comprehensive plan is innovative in its holistic commitment to principles of sustainability, and, if implemented fully, will help position the community as a leader in green building, energy efficiency, and other climate-friendly techniques. One other outstanding aspect of the plan is the high level of community outreach, education, and involvement that was part of its production. This involvement was the key to wide acceptance of the sustainability concepts. Although rebuilding an entire community from scratch is an uncommon event, achieving buy-in from multiple stakeholders is something that is always sought but often elusive in the plan-making process.

Within days of the storm, Greensburg citizens, city staff, and elected officials were talking about the rebuilding process. The Federal Emergency Management Agency (FEMA) erected a large circus tent in the city park, and it was in this makeshift community headquarters that planning meetings, community meals, and critical decisions were made. By bringing the planning process into this communal space, thinking about a future for the town became one way residents dealt with the trauma of the storm. They were assisted in their work by state and federal

nizations and individuals. So they planned carefully, initially forming a core planning team led by Greensburg city staff. BNIM Architects and architect John Picard were retained in October 2007. BNIM led efforts to facilitate community participation, conduct design analysis, coordinate planning and design recommendations, and distill the team's findings into a legible plan. Picard served as a lead advisor to the city. The core planning team received additional assistance from numerous state and federal agencies and departments, including FEMA, the U.S. Environmental Protection Agency, the National Renewable Energy Laboratory, the U.S. Department of Agriculture Office of Rural Development, the Kansas Corporation Commission, the Kansas Energy Office, the Kansas Housing Resources Corporation, and the governor's office. The nonprofit Greensburg GreenTown soon emerged to serve as an important grassroots organization that educates and excites the community about the ideas of rebuilding green.

City officials and their partners also knew that if they could involve as many Greensburg residents as possible in the process and keep them well-informed, they would create a large and supportive constituency for the goals contained in

(continued on page 59)

(continued from page 58)

the comprehensive plan. To that end, immediately after the storm Governor Kathleen Sebelius asked Kansas Communities LLC, led by Terry Woodbury, to facilitate a "Public Square" process to support long-term rebuilding and develop citizen leadership in Greensburg. The Public Square Steering Committee was established, with four sectors representing different aspects of the community: government, business, education, and health and human services. Some members of the steering committee were recruited into a Recovery Action Team that met even more frequently with other members of the core planning team. They were tapped to discuss portions of the plan and help the team communicate with the broader community. On several occasions, Greensburg's high school students influenced the direction of the rebuilding effort. A group of students embedded themselves in the planning process and directly contributed ideas related to the location of the future school, the commitment to sustainability, and even the decision to rebuild. All of these efforts were coordinated with the Kiowa County Business Redevelopment Group, city staff, Greensburg GreenTown, the Planning Commission, and the City Council. Every meeting and discussion about rebuilding dealt directly with the effort to become an economically, environmentally, and culturally sustainable community.

as we were and continue to die?" Gurnee continues: "There was another facet to choosing green in western Kansas. Rain barrels . . . windmills . . . natural light . . . attention to prevailing winter winds . . . solar orientation . . . and especially being 'stewards of the land': these are things granddaddy and great-grandpa cared about when they homesteaded western Kansas. Becoming a sustainable community was going back to the citizens' roots for many."

To be sure, there has been a learning curve. Gurnee tells the story of a motel at the edge of town that received only minor damage. To go "green," the owner replaced the roof with green shingles and painted the trim green. That was before the comprehensive plan and the public meetings on it. Most everyone understands and buys into the concept now. From outward appearances, a tax credit–financed apartment complex looks like every other one in the region. However, it was recently certified LEED Platinum.

There are some naysayers. There is a move afoot to relax some suggested standards in the comprehensive plan. Business has been slow to come to town. Sustainable building is perceived to be prohibitively expensive. Gurnee feels that the primary reason for slow business development is the state of the national economy. At trade shows, many groups show an interest in

(Left to right) **Figure 5.6.** *City Administrator Steve Hewitt addresses the audience at the comprehensive plan public meeting, December 20, 2007;* **Figure 5.7.** *GREEN club members address the audience at the comprehensive plan public meeting, December 20, 2007;* **Figure 5.8.** *Wind turbines provide auxiliary power to the LEED Platinum 5.4.7 Arts Center, Greensburg, Kansas.*

Two years after the tornado, Greensburg Community Development Director J. Michael Gurnee, AICP, reports that the implementation of the comprehensive plan is on target. The city hall, hospital, and school are being rebuilt to LEED Platinum standards. The courthouse was retrofitted to LEED Gold standards without harming the historic fabric of the 1914 structure. Additionally, the streetscape project along Main Street is underway. This includes the creation of rain gardens, use of irrigation from reclaimed water, and installation of street furniture made from reclaimed materials. The entire town has LED streetlights.

Gurnee is working on a new land development code to match the sustainability vision and goals in the comprehensive plan. He says, "The most difficult part of putting this together is to make the document workable for a small town that will not always have a professional planning staff. The codes have to be so clear that a paraprofessional can handle them." Gurnee expects to have the project finished before the end of 2009.

Gurnee discussed the process that helped the citizens choose a sustainable vision. Like many small towns on the high plains, Greensburg was a stagnant, if not dying, community. He recalls that former mayor John Janssen often said, "We were a dying town before the storm; do you want to rebuild exactly

All photos by J. Michael Gurnee, AICP, Greensburg Community Development Director

Greensburg but are having difficulty getting building loans. As Gurnee says, sustainable building is only expensive if it is not well thought out and planned for. Despite some of these setbacks, the City Council recently made it clear that it intends to maintain the plan.

In the face of unimaginable devastation, the people of Greensburg worked with one another, their city government, and outside agencies to create an innovative and groundbreaking sustainable comprehensive plan that will help them re-create their community in a sustainable, climate-friendly way over the years to come. ◀

RETHINKING PLANNING PRACTICE

Global challenges require new ideas and new approaches. While the strategic points of intervention outline opportunities in the traditional planning process, planners, planning commissioners, and others involved in the planning process should explore new ways to create changes in their community to reduce energy use and GHG emissions. Despite the daunting picture of what the future might bring if global temperatures continue to rise, planners have significant opportunities to confront this challenge and effect changes that may reduce the negative impacts of climate change.

CHAPTER 6

Development Patterns

 As every planner and real estate agent knows, location matters. This chapter explores the multiple implications of location and development patterns for energy efficiency, greenhouse gas reductions, and effective climate adaptation. Land-use planning strategies discussed in this chapter focus on three main areas:

- The Energy and Climate Benefits of Compact Mixed Use Development;
- Concentration of Development in Low-Risk, Low-Sensitivity Areas; and
- Population Migration Associated with Climate Change.

ENERGY AND CLIMATE BENEFITS OF COMPACT MIXED USE DEVELOPMENT

Reducing energy use and greenhouse gas emissions has implications for both where and how we build. Location, density, proximity, connectivity, and diversification of land uses can all influence energy use and greenhouse gas emissions. Promoting compact development—which is characterized as a mix of uses at medium to high densities with centered development, interconnected streets, and pedestrian and transit-friendly design (Ewing et al. 2008)—is an important strategy that communities can use to achieve energy and climate goals.

Planners have an important ally in the real estate market. The demand for more compact residential development is growing as a result of demographic trends toward smaller household sizes and greater diversity of housing needs, along with shifting preferences for in-town, walkable neighborhoods and smaller houses. Figure 6.1 quantifies this demand by residential land-use category. Demand for attached and small-lot residential development is expected to increase, while demand for large-lot residential development appears to have peaked. While there will be regional differences with regard to the demand for these housing types, there are widespread opportunities to promote more compact residential development.

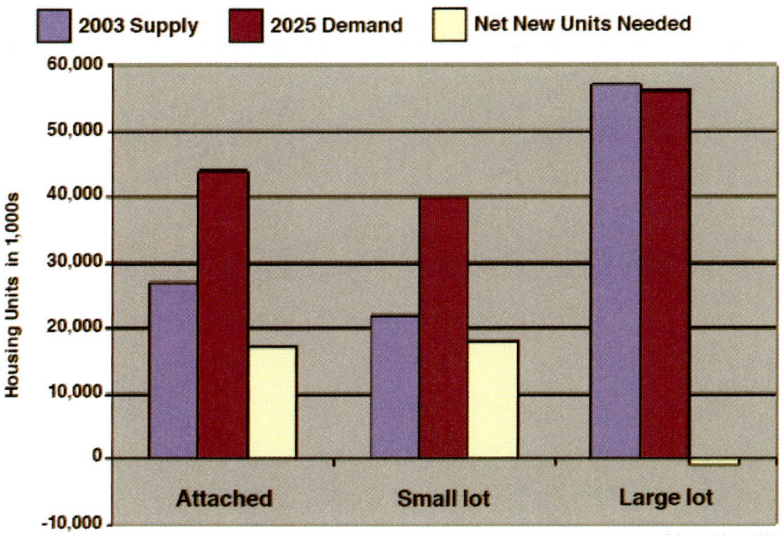

Figure 6.1. Anticipated demand for various kinds of lots

Adapted from Nelson 2006

Compact mixed use development patterns consistent with "smart growth" principles inherently have many potential energy saving and GHG reduction benefits, but successful integration of energy and climate strategies into local development patterns will require considerable thought and attention. Planners and local communities will need to evaluate and integrate energy and climate considerations more explicitly and more rigorously than has been typically required in the past. The potential energy and climate benefits related to development patterns include:

- More efficient transportation systems
- More efficient heating and cooling of buildings
- Enhanced opportunities for renewable energy
- Avoided land clearing and carbon storage losses
- Reduced infrastructure costs and energy use
- Reduced urban heat island effects

More Efficient Transportation

Transportation benefits of compact mixed use development are familiar to most planners: having everyday destinations closer together means fewer miles traveled by motor vehicles, more opportunities to walk or bike, and greater viability of public transportation.

Figure 6.2. Sprawling development patterns

Several studies have shown that household travel demand is strongly influenced by where that household is located. An analysis by the Jonathan Rose Companies (see McIlwain 2008) indicates an average household living in a typical suburban house will use a total of 240 million Btus (MBtus) per year, slightly more than half of which is for transportation. A suburban household in a suburban home designed and constructed with green building practices will consume an estimated 164 MBtus annually. In an urban setting, the average household is estimated to consume 143 MBtus, and a household in a green home consumes 89 MBtus, with the reductions due largely to decreased transportation energy use.

The relative impact of transportation/land-use policies can also be seen (e.g., Wilson and Navaro 2007) by comparing the amount of energy used to heat and cool an office building constructed with either green building or conventional design with the amount used in the daily commutes of the people who work there. For a conventional suburban building whose workers commute the average national distance of just over 12 miles, the energy used for commutes exceeds the energy used to operate the building by more than 30 percent. If the building is constructed with green building techniques, that figure rises to 137 percent. Thus, transportation can be the dominant consideration for planners attempting to reduce energy usage as a strategy for affecting GHG emissions.

In studies of metropolitan areas, regional accessibility—a measure of how connected development locations are to other important destinations in the region through proximity and one or more travel modes—emerges as a paramount factor. Residents of highly accessible areas tend to have measurably lower energy and GHG footprints per capita because of their transportation options and efficiency.

In metro areas with multiple employment centers, strategies should focus on providing and enhancing transit options between employment centers for dual-income families and employees who may travel frequently between regional offices. Facilitating adequate housing opportunities and schools near employment centers can also help reduce vehicle miles traveled.

Figure 6.3. Some Seattle bicycle commuters take the ferry as part of their trip.

Regional scenario planning—pioneered in places like Sacramento, California—is emerging as a promising tool for addressing energy use and GHG emissions from transportation. Scenario analysis is simply the comparison of different options for accommodating projected development trends—new residents, housing, businesses, and so on—though this may require sophisticated data collection and analysis tools. A full scenario analysis not only can show where and how new land uses will be accommodated but can estimate the infrastructure investments, operating costs, energy use, and GHG emissions associated with different scenarios. The preferred scenario developed by Sacramento, which followed smart growth principles, was estimated to reduce GHG emissions by 15 percent over the business-as-usual or "base case" scenario. (See the Sacramento case study.)

On a national scale, one study estimates that if 60 percent of new growth occurred in compact, mixed use developments, the volume of carbon gas emissions that reach the atmosphere each year could be reduced by up to 85 million metric tons by 2030, which is comparable to a 28 percent increase in passenger-vehicle fuel efficiency (Ewing et al. 2008).

More Efficient Heating and Cooling of Buildings

Compact development patterns also tend to increase the proportion of attached and multi-unit buildings, both residential and commercial. Shared walls, as well as generally smaller unit sizes, create opportunities for lower energy use per capita. Other factors being equal, attached and multi-unit housing units tend to use significantly less energy for heating and cooling compared to detached houses—by 20 percent or more, according to some estimates (Ewing and Rong 2008; EIA 2005). In practice, however, other factors affecting energy use for heating and cooling buildings—such as building design and construction, efficiency of heating and cooling systems, household income, and so on—are generally not equal.

Methodologies for estimating the energy use of different building types have been developed (see Oak Ridge National Laboratory n.d.). Such information can be applied to compare different building types and configurations and to provide a sense of relative potential energy savings for heating and cooling attributable to density. Approximate values can be developed for a given area to compare large-scale and long-term development scenarios.

Enhanced Opportunities for Renewable Energy

Compact development can provide inherent energy efficiencies, but planners must also grapple with how and where energy is generated, transmitted, and distributed. Compact development generally enhances opportunities to deploy different types of renewable energy or energy efficiency–enhancing technologies. The cost of infrastructure to carry electricity from a renewable source is often a critical factor in determining its feasibility. The proximity of energy-demanding land uses created through compact development can increase the cost-effectiveness of installing new transmission and distribution infrastructure.

The infrastructure required for renewable and alternative energy sources may also include pipelines for natural gas and biomethane, steam pipes for transmitting thermal energy from biomass-fired district heating and cooling facilities, or pipes for distributing solar hot water and water heated or cooled by geothermal sources. The per unit costs of these types of infrastructure can be prohibitively high for low-density development but are more viable where costs can be shared among more units at higher densities. Most of the community energy strategies that can aid transition

Figure 6.4. Multifamily housing

SACRAMENTO AREA COUNCIL OF GOVERNMENTS (SACOG) BLUEPRINT LAND-USE STUDY

Population: 1,948,777

In 2002, the Sacramento Area Council of Governments (SACOG), the regional planning body for a six-county region that also serves as a federally designated metropolitan planning organization, initiated a two-year project to examine present land-use patterns and future growth scenarios. The goal of the project was to determine how the region could grow in a smart and sustainable manner through 2050. Community outreach and inclusion were central themes of the process.

The Blueprint process began when a base case study was released showing how the Sacramento region would look in 2050 with unchecked growth. Research done by the Center for Continuing Study of the California Economy estimated 1.7 million more people in the region by 2050; the number of homes was expected to more than double. Realizing that this unprecedented growth could not be accommodated by current infrastructure, SACOG set out to comprehensively plan for the future in a way that would minimize the negative impacts of urban sprawl and avoid increased traffic congestion.

The Blueprint project used state-of-the-art modeling tools in conjunction with an innovative community-outreach strategy to determine how best to plan for the Sacramento region's growth. SACOG developed planning tools to examine the effects of land-use patterns on transportation, air quality, and the economy. Information gathered from the Blueprint regional-planning process was used to revise local plans and shape future urban growth and transportation patterns in conjunction with smart growth goals and community opinions and values.

Figure 6.5. Participants at a SACOG Blueprint workshop

SACOG partnered with Valley Vision, a nonprofit organization, to develop a community-outreach strategy that brought together community members, elected officials, and business leaders for a series of surveys, regional workshops, and conferences. SACOG used PLACE³S (PLAnning for Community Energy, Economic and Environmental Sustainability)—software that combines planning and design with quantitative analysis—to produce scenario maps, tables, and charts that demonstrated how the region would grow under various land-use policies, including policies currently in place. Planners then worked with community members to create alternative scenarios for future growth.

The initial budget for the project was $500,000, though the actual realized cost was in the low millions. The majority of financing was used for both development and modification of tools and the planning and conducting of community workshops. Other municipalities interested in using the Blueprint process can benefit from already-developed software tools, including PLACE³S.

(continued on page 66)

(continued from page 65)

The Preferred Blueprint Scenario, adopted in 2004, was the end result of a year-long series of workshops that included more than 30 neighborhood-scale workshops held in all parts of the Sacramento region. Local government staff, elected officials, and more than 1,000 citizens were given the opportunity to express their preferences regarding how the region should grow. Planners from across the region worked together to manage input from the workshops and develop a preferred scenario consistent with the smart growth and sustainability principles of the Blueprint project. In addition to promising measurable VMT reductions per household, the preferred scenario promotes compact mixed use development and increased transit choices as alternatives to low-density development. ◀

to renewable, low-carbon energy alternatives (see Chapter 2) are enhanced and made more feasible by density and the proximity of energy users. There are also efficiency losses associated with sending electricity over longer distances. High voltage helps minimize such losses for central transmission lines that may traverse very long distances, but the losses are more significant over the thousands of miles of low-voltage lines that distribute power to individual end uses from either a local source or the nearest transmission substation.

Avoided Land Clearing and Carbon Storage Losses

The development of land that was previously in agricultural use, forested, or natural vegetative land cover creates a net addition of GHG emissions because that land is no longer able to sequester and store carbon in its soil and vegetation (see Chapter 11). The disturbance of soil alone makes a large one-time contribution to atmospheric CO_2. This is one more factor that can be integrated into the analysis of the net GHG emission impact of any development scenario, whether project level, local, or regional.

Figure 6.6. Suburban development on farmland

Reduced Infrastructure Costs and Energy Use

Development patterns have consequences not only for infrastructure directly associated with energy but also for the energy use associated with water, transportation, and other types of infrastructure. The manufacture, construction, maintenance, and operation of every linear mile of pavement or pipeline involve a substantial amount of indirect or "embedded" energy. Approximately 4 percent of U.S. electricity consumption goes to moving and treating water and wastewater (EPRI 2002).

Measuring embedded energy on a project-by-project basis using general rules of thumb can give planners and local communities a sense of the potential significance of this effect and the additional energy benefits of compact development. Further, including dollar costs as well as energy savings in such analyses can generate political and developer support for compact development.

Reduced Urban Heat Island Effects

The relative thermal efficiency of compact development comes with a caveat. Increased density can amplify the phenomenon known as the "urban heat island" effect, whereby the concentration of pavement and roofs in urbanized areas increases the thermal mass of materials that absorb and store heat from the sun. The heat island effect is exacerbated by the black or dark-colored surfaces that prevail in many cities and towns—such surfaces tend to absorb a greater proportion of the energy coming from the sun than lighter-colored surfaces do.

The urban heat island effect increases ambient temperatures in urbanized areas, which can add to the average summer cooling load. Peak summer temperatures can be as much as eight degrees higher in city centers than in the surrounding countryside (DOE 1996). In winter, the warming effect can help offset heating loads, but this effect. is less significant because the sun's energy is less intense, the sun rises less high in the sky, and in some areas snow cover increases reflectance.

Measures to address this direct warming and help "cool" buildings are discussed further in Chapters 10 and 11. But in terms of development patterns, the primary strategy is to minimize excessive paved areas and maximize vegetated areas despite the increased density. Conservation of natural areas, parkland, gardens, and other open space within and throughout areas with higher development density have been shown to have a significant effect on ambient temperatures in central areas.

Implementing Compact Development

Since energy and climate change considerations provide extraordinary reasons for pursuing compact development patterns, how can planners most effectively pursue implementation of those patterns? Planners must first recognize the variety of reasons why compact development patterns may be successful in their communities. Economic changes, including the transformation of the U.S. economy from a manufacturing-based model to a service-based one, support the placement of businesses in close proximity to residences, since operational impacts such as noise and emissions that are associated with service-based industries are significantly less intrusive than those associated with manufacturing operations. In addition, lengthening commutes and inadequate transportation networks are causing workers to rethink their residential location choices, further strengthening the market for urban-style housing close to centers of employment. Rising energy costs will reinforce these housing and transportation preferences.

There are a wide range of tools and techniques that planners can use to implement a more compact development pattern. These include:

- identifying and utilizing infill and brownfield sites for development;

- encouraging adaptive reuse of buildings that are functionally obsolete (for example, converting a functionally obsolete industrial building into residential units with ground-floor retail; see Chapter 10 for a discussion of the embedded-energy benefits of such reuse);

- revising codes and standards to allow small-lot, mixed use, and cluster development;

- facilitating transit-oriented development in areas served by extensive public transportation;

- employing urban growth or service boundaries, which limit where development may occur or where urban services such as water and sewer may be provided.

CONCENTRATION OF DEVELOPMENT IN LOW-RISK, LOW-SENSITIVITY AREAS

Climate change is expected to create or expand areas at risk from flooding, storm surges, wildfires, and other weather-related disasters. It may also create more urgency about the need to protect sensitive resource areas such as groundwater recharge areas, river and stream corridors, endangered species habitats, and natural habitat areas. Too often, the need to limit new development in high-risk areas and focus investment in low-risk areas is overlooked.

Limiting development typically does not mean sudden or total disinvestment in hazard-prone areas that already contain established communities and neighborhoods. For example, a community may establish buy-out or easement programs that allow for the gradual relocation of persons in areas facing shoreline erosion or repetitive flood loss. Areas facing more frequent wildfire threats can implement site design and event response techniques that can improve their resilience. Coastal wetland and dune restoration can create buffers that afford population centers greater protection from storm surge. Conservation measures can be taken that promote the more efficient use of water resources. However, in many cases a long-term plan to ultimately shift development to more suitable areas may represent the most prudent and sustainable response to these challenges.

Incentives, design techniques, and regulatory approaches that promote denser development in low-risk areas are especially effective measures to improve community resilience against the effects of climate change and

Figure 6.7. A frequently flooded residential development, Iowa

energy-related vulnerabilities. Three additional areas of emphasis for planners wishing to incorporate strategies that direct development away from high-risk areas are discussed below.

Floodplain Management and Erosion Control

As the climate changes, many areas of the United States, even those that will be relatively drier, may experience precipitation events that produce greater amounts of precipitation over shorter durations. Whether this precipitation is snow or rain, it will create the potential for enhanced runoff and greater flooding. Adapting to these circumstances may require communities to reevaluate existing regulations governing floodplain and stormwater management and erosion control. Communities with recently updated Flood Insurance Rate Maps (FIRMs) may need to review the as-

Figure 6.8. Soil erosion near residential development

sumptions used in establishing flood-classification boundaries to ensure that potential climate change impacts have been accounted for. Communities that have certified local flood-management programs but lack recently updated FIRMs might consider increasing base elevation requirements to minimize potential flood damages. Areas that have development on steep slopes will have to evaluate landslide and mudslide potentials and address them through design standards, density limitations, transfers of development rights, conservation measures, and other mechanisms. In areas experiencing gradual erosion due to sea-level rise, a "rolling easement" approach may be adopted to allow the gradual retreat of existing development in lieu of shoreline hardening.

Infrastructure

Decisions about infrastructure—initial installation, replacement, and maintenance—need to incorporate an awareness of potential risks from climate change and from energy issues such as availability and efficiency. In the past, infrastructure decisions such as the provision of roads and utilities on barrier islands and major water reservoirs in the arid southwest have directly supported the migration of populations to areas that face risks due to climate change and energy issues. Identifying population and infrastructure vulnerability and ensuring a rational relationship between the installation and maintenance of infrastructure and the associated human and capital risks are critical tasks for planners. In addition to decisions about where to place and replace infrastructure, decisions about infrastructure design must also be made carefully. Many flood-protection approaches have emphasized structural design concerns. The limitations of this approach became apparent after Hurricane Katrina. Structural failures resulted in severe flooding of large portions of New Orleans, but more attention to nonstructural approaches—which may include relocation of vulnerable populations and structures, floodproofing, utilization of natural systems to direct or divert floodwaters, planning measures to direct growth to less vulnerable areas, conservation easements, and similar mechanisms—might have better mitigated the damage.

Conservation Approaches

In many cases, areas that are vulnerable to flooding, erosion, storm surge, and other climate change threats have scenic value that contributes positively to community character. Similarly, areas that are difficult to serve efficiently with utilities, such as rural areas, hillsides and mountains with steep slopes, and deserts, are also often scenic. Focusing conservation efforts on highly vulnerable locations can consequently not only minimize population risk but also form the basis for economic development and efficient infrastructure strategies. In many cases, conservation of vulnerable areas can also provide protective buffers, which allow protection for better-situated population centers. (See the Soldiers Grove case study.)

POPULATION MIGRATION ASSOCIATED WITH CLIMATE CHANGE

People have always migrated because of climate, but communication advances combined with the globalization of labor and transportation have made some sectors of the U.S. population highly mobile and thus able to take advantage of economic opportunities, desired lifestyles, and favorable climate conditions. This widespread mobility intensifies a significant adaptive climate change/land-use challenge for local land-use programs and policies: how to accommodate the effects of people migrating to a community because of a favorable climate, and how to respond when they leave because of an unfavorable climate.

Some climate change impacts may be quite severe and essentially permanent, such as inundation of coastal land areas and the constrained availability of potable water due to drought conditions or saltwater intrusion. Other impacts may be severe, event-oriented, and potentially serial, such as stronger hurricanes, floods, wildfires, and heat waves. Any one of these impacts has the potential to result in population migration to more favorable locales. For example, a portion of the population of New Orleans chose not to return to that city after Hurricane Katrina.

In areas experiencing net population growth due to climate change, the principal challenge will be managing this growth. Areas experiencing net population loss due to climate change will need to face the larger question of the extent of the impact. Severe and essentially permanent climate change effects, such as inundation, can result in the relocation of entire communities. Other communities may find themselves constrained by limits on resources, such as water, or by higher temperatures and rainfall patterns, which make them less desirable places to live or visit. Once these vulnerabilities are identified, measures can be taken to manage the effects of population loss.

Managing Climate Change In-Migration

Adapting to population gains associated with climate change is just another form of growth management, but with some subtle distinctions. People and businesses relocating as a result of climate change will be relatively more attracted to communities that demonstrate climate change resilience; in general, they will not want to substitute one type of climate vulnerability for another. Consequently, communities interested in attracting and accommodating climate change–related growth will want to create mitigation and adaptation plans and programs that respond to their local and regional climate change issues, including water supply, floodplain management, disaster preparedness, and effective growth management. Additionally, such communities may want to consider customized social-services outreach to these new residents, especially if there are substantial numbers of people relocating due to traumatic events. In some cases, in-migration may be the result of humanitarian efforts to relocate threatened populations from other nations, in which case such outreach programs become both more

SOLDIERS GROVE, WISCONSIN
Population: 605

Many communities face the potential of increased flooding risks due to climate change. Anticipated increases in precipitation in the northeastern and midwestern United States may require communities there to learn to adapt to unavoidable changes.

In the past, communities have adapted to the changing environment. Soldiers Grove, Wisconsin, is just one example of a community facing an environmental challenge head-on and adapting to the inevitable impacts. In 1978, town leaders decided to completely shift development away from a high-risk floodplain area.

The history of Soldiers Grove is intrinsically bound to the Kickapoo River. The town experienced its first major Kickapoo flood in 1907. Record-breaking floods followed, leading up to 1937 when the town petitioned Congress for assistance with a flood-control project. Plans to upgrade stormwater infrastructure were hampered by the lack of funding and other obstacles, barriers that communities still face today.

Compounding the environmental costs of flooding were devastating economic costs to the community. A state-required floodplain zoning ordinance prohibited all new construction in the central business district and placed limitations on maintenance to existing buildings. When a final flood-control plan was released by the U.S. Army Corps of Engineers in 1975, with a dam and levee as keystones, it became clear to community leaders that the proposed levee would not be adequate. The infrastructure upgrades would protect only about $1 million worth of property, cost more than double that amount to build and maintain, and do nothing to solve the economic crisis the town faced.

Figure 6.9. Flooded streets in Soldiers Grove, Wisconsin

The town devised a radical solution to its problem: evacuate the town and rebuild out of the floodplain. With modest financial resources and no state government support, the town worked with a team of landscape architects to conduct a feasibility analysis and began to extend utility services to designated relocation sites. A devastating flood event in 1978 provided the needed catalyst for the federal government to support relocation. Following that flood, the U.S. Department of Housing and Urban Development granted the village $900,000 to begin relocating to lower-risk areas.

(continued on page 72)

(continued from page 71)

The town treated the move uphill, completed in 1983, as an opportunity not only to adapt to its flooding problems but also to mitigate future environmental issues. Inspired by the energy crisis of the 1970s, town leaders pledged to make all buildings in the new town center energy efficient and solar heated. A community-wide solar-access law was passed, and a new ordinance required all new commercial buildings to receive at least half their heating energy from the sun. The changes spurred new life in the town economy. Construction brought new jobs, and the solar regulations, unique at the time, brought increased tourism to America's first "solar town."

Any doubt about the necessity of the 1978 move was removed in 2008, when another major flood event struck the town. Damage was concentrated around Soldiers Grove Park, an area that stands on the town's former Main Street. The new commercial district was mostly untouched.

Communities facing impacts brought on by climate change can learn from Soldiers Grove's dedication to solving the problem and willingness to adapt, recognizing that future conditions may be very different than those of the past. Although moving the town was a significant investment, *not* moving the town would likely have resulted in even greater costs. In assessing long-term environmental impacts, it is critical to make wise choices in adapting development patterns to account for probable climate change impacts. ◂

complicated and more necessary in order to address cultural differences and unique social-services needs.

Growth management techniques may include establishing a compact development pattern, creating an interconnected multimodal transportation network, effectively addressing ecosystem impacts, implementing green building and site design programs, enhancing disaster management planning, ensuring the availability of adequate infrastructure, insulating economic development activities from climate change impacts, and addressing potential public health issues.

Managing Climate Change Out-Migration

Out-migration effects are generally less familiar to planners. Some individual large cities (such as Philadelphia and Richmond) have seen population losses, but even these are located in growing metropolitan regions. Still, we can learn from these cities' experiences in adapting to declining populations.

While out-migration creates a number of challenges (declining tax base, constraints on economic development, public safety issues associated with property abandonment, and so on), it also offers opportunities to improve housing affordability, adaptively reuse abandoned buildings and sites, address traffic congestion, introduce new open space and parkland into urban areas, create partnerships with businesses and institutions that remain in the community, and unite the human capital of the community in positive ways.

Extended disinvestment scenarios will require planners to secure the assets of a community or neighborhood in a way that maintains or potentially enhances the overall quality of life as determined by the residents. These techniques can include tapping institutional resources for assistance (e.g., employer-supported housing programs that revitalize surrounding neighborhoods), expanding open space by selected reclamation of abandoned properties, and converting vacant lots to community gardens or parking. Flexibility and creativity will be essential skills for planners in areas experiencing out-migration due to climate change.

TOOLS FOR GUIDING DEVELOPMENT PATTERNS

Visioning and Goal Setting	Review the community's energy and climate goals and consider what future development patterns might look like
Plan Making	Review land-use maps and policies to see if they allow compact development, identify areas where climate risks are minimized, and plan for anticipated future population growth or decline
Standards, Policies, and Incentives	Establish policies and standards for making energy, GHG savings, and climate resilience factors in planning and development decisions
Development Work	Encourage compact development in low climate-risk areas
Public Investment	Prioritize public investments that promote compact development patterns and avoid areas subject to flooding and other climate-related risks

CHAPTER 7

Infrastructure and Utilities

Infrastructure decisions are central to responses to new energy and climate challenges. Climate change can pose threats to existing infrastructure and alter the design and maintenance of new infrastructure. Infrastructure decisions also shape energy use, while addressing energy infrastructure issues, in particular, is implicit in transitioning to renewable energy sources. This chapter addresses the major energy and climate concerns facing infrastructure and utilities including:
- Risk-based Infrastructure Planning and Design
- Transportation Infrastructure
- Water and Wastewater Infrastructure
- Energy Infrastructure

RISK-BASED INFRASTRUCTURE PLANNING AND DESIGN

A significant amount of the public infrastructure in the United States may face risk of physical damage due to climate change. Increased flooding potential and sea-level rise may threaten sewage treatment facilities that have been constructed at the water's edge for gravity collection and discharge purposes. Water reservoirs may have reduced supply due to drought and increased evaporation resulting from higher temperatures. Airport runways may require lengthening due to higher temperatures and humidity, which reduce air density, affecting aircraft performance. Stormwater management systems may be overwhelmed by increased intensity in precipitation events. Flooding, inundation, and other extreme events may threaten infrastructure to the point that it needs to be relocated or abandoned. Most infrastructure facilities are intended to have relatively long lives; this longevity translates into a greater risk factor for future damage as climate changes over time.

Figure 7.1. Flooding in New Orleans after Hurricane Katrina

Consequently, infrastructure decisions should take into account potential risks in siting, budgeting, capital improvement programming, maintenance, standards and specifications, and development regulations. There are a variety of resources available to planners in assessing risks to infrastructure. These include:

- State climate plans
- Regional climate centers
- State climatologists
- Research universities
- Federal agencies

Most infrastructure represents long-term capital investment. New construction will need to take climate change impacts into account as part of the design process. Retrofitting existing infrastructure will require a comprehensive assessment of the costs and benefits of addressing the effects of climate change. For example, decisions about bridge replacement in areas that will flood more frequently or face inundation threats need to take into consideration the climate vulnerability of the roads accessing the bridges and the vulnerability of the areas being accessed; if these roads and areas are also vulnerable, addressing bridge elevation deficiencies hardly solves the overall problem.

Sea-level Rise Issues

Inundation represents a significant threat to coastal area infrastructure. Low-lying coastal areas, especially in regions that are experiencing subsidence, may experience permanent or frequent inundation due to sea-level rise. Infrastructure in these areas is highly vulnerable to damage or to complete obsolescence and may prove increasingly inadequate in evacuation circumstances. Sea-level rise may also intensify the effects of storm surge.

Figure 7.2 on page 76 illustrates the potential effect of an approximately half-meter (48.5 cm) sea-level rise on coastal North Carolina, one of the states that is more vulnerable to increases in sea level. In this scenario, more than 1,000,000 acres of low-lying land would be inundated. Clearly, in areas facing regular or periodic inundation, maintenance and replacement schedules or investments in new infrastructure must be evaluated with these risks in mind.

Precipitation Issues

Stronger storms and more intense precipitation events create risks to infrastructure and infrastructure operations. These risks include facility damage from flooding, storm surge, and landslides and mudslides. Weather-related accidents will likely increase in frequency. Overall snowfall may decrease, but the amount of snow in an individual event may become more significant. In some areas, ice storms may become more frequent.

Communities in regions where such precipitation changes can be anticipated will need to account for related issues. Responses include an increased emphasis on stormwater management and flood protection as part of infrastructure design, incident response programs geared toward heavier precipitation and stronger storms, and adjustments to manpower and resources associated with snow and ice removal.

CHICAGO: CLIMATE ACTION PLAN
Population: 2,836,658

The Chicago Climate Action Plan focuses on implementation, is grounded in research, and looks at potential adaptation and mitigation measures.

Innovation

The City of Chicago made a commitment to research, education, and action to support both mitigation of greenhouse gases and adaptation to impacts of climate change. The approach to climate action undertaken by the city is notable in that it identifies not only the potential impacts of a changing climate but also the risks associated with those impacts, thus connecting climate change predictions with economic impacts.

To strengthen the adaptation planning component of the Chicago Climate Action Plan and promote a better understanding of the issues, potential impacts, and solutions, a Climate Change Task Force convened experts from the University of Illinois (UIUC), Texas Tech University (TTU), and Oliver Wyman and Associates, a consulting group specializing in corporate risk. The resulting climate change impact reports "Climate Change and Chicago, Projections and Potential Impacts" (Hayhoe et al. n.d.) and "Economic Impact Analysis of Climate Change for the City of Chicago" measure the likelihood of predicted impacts and the probable consequences for or risks to the city's water, health, ecosystems, and infrastructure. By measuring the potential cost of climate impacts on the city's infrastructure and quantifying the monetary value of action versus that of inaction, Chicago was able to identify the most critical areas in which to focus adaptation strategies.

The Process

At the request of Chicago mayor Richard Daley, the Chicago Department of the Environment launched a climate change initiative in late 2006. The initiative was led by the Climate Change Taskforce, consisting of representatives from business, civil society, government, and labor who were tasked with developing a climate action plan for the Chicago metropolitan region. The city commissioned climate scientists, engineers, and risk experts to conduct research, develop a methodology, and recommend actions to mitigate and adapt for a changing climate. Multidepartmental working groups that included representatives of the city departments of planning, transportation, and water management were involved in the plan making process.

The Plan

The Chicago plan was adopted in September 2008. Baseline greenhouse gas emissions and quantitative analysis of potential emissions reductions are detailed in it. The action plan itself focuses on implementation of strategies in five areas to meet the goal of reducing GHG emissions to 25 percent below 1990 levels by 2020 and to 80 percent below by 2050. The plan proposes 26 actions to reach reduction goals including adaptation, construction of green and energy efficient buildings, use of renewable energy, transportation system improvements, and reductions in waste and industrial pollution.

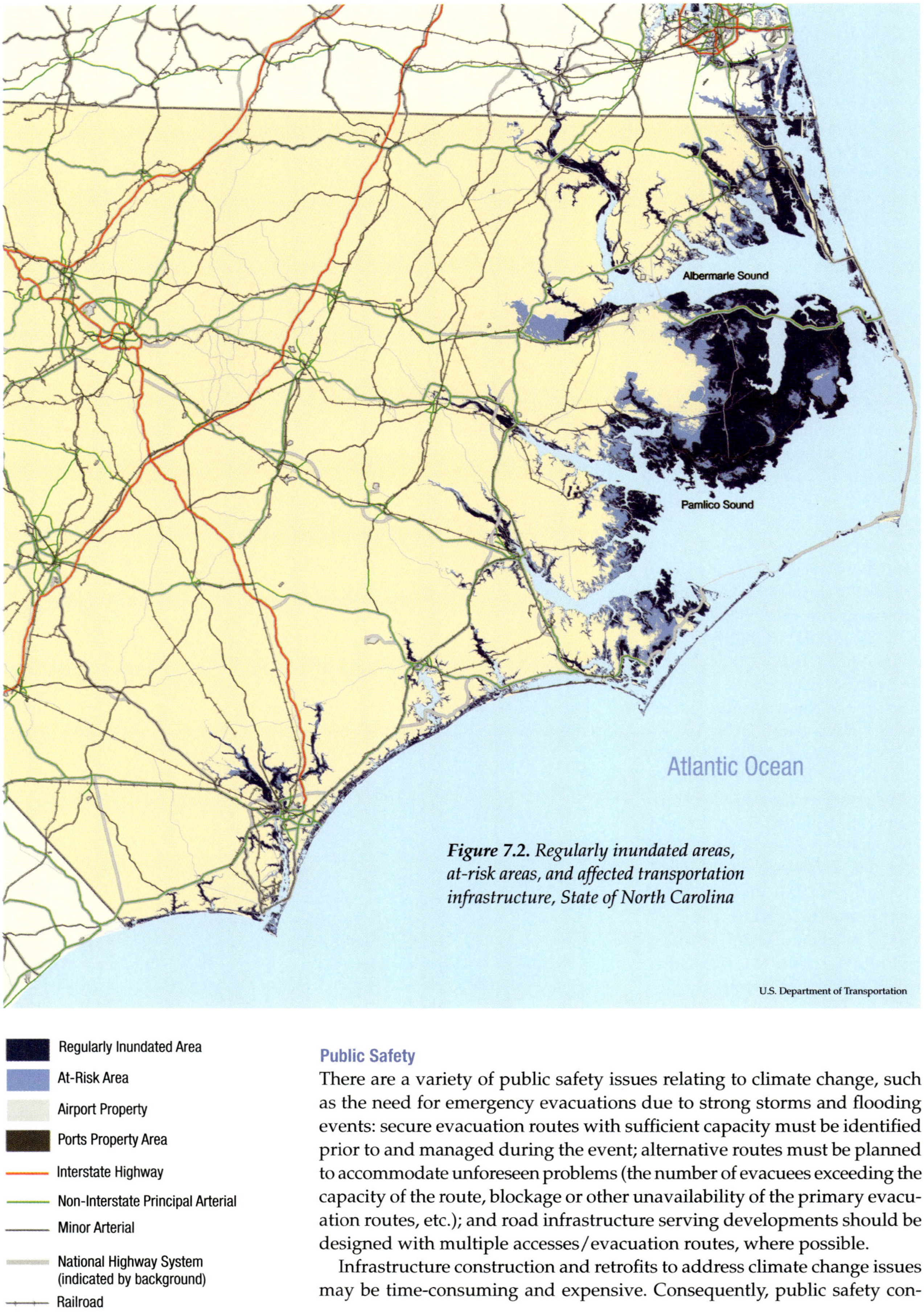
Albermarle Sound
Pamlico Sound
Atlantic Ocean
Regularly Inundated Area
At-Risk Area
Airport Property
Ports Property Area
Interstate Highway
Non-Interstate Principal Arterial
Minor Arterial
National Highway System (indicated by background)
Railroad

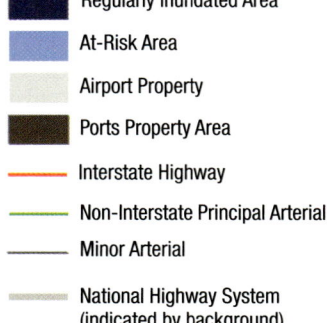

Figure 7.2. *Regularly inundated areas, at-risk areas, and affected transportation infrastructure, State of North Carolina*

U.S. Department of Transportation

Public Safety

There are a variety of public safety issues relating to climate change, such as the need for emergency evacuations due to strong storms and flooding events: secure evacuation routes with sufficient capacity must be identified prior to and managed during the event; alternative routes must be planned to accommodate unforeseen problems (the number of evacuees exceeding the capacity of the route, blockage or other unavailability of the primary evacuation routes, etc.); and road infrastructure serving developments should be designed with multiple accesses/evacuation routes, where possible.

Infrastructure construction and retrofits to address climate change issues may be time-consuming and expensive. Consequently, public safety concerns should be an integral part of risk-based decision making about infrastructure design and investments. Examples include:

- Installing permanent barriers or gates that can be activated during flood events to block access to frequently flooded roadways.
- Creating emergency access points to developments with limited accessibility during flood or storm surge circumstances.

Due to the potential for heavier precipitation events, public safety operations may need to respond to increased numbers of weather-related accidents involving transportation facilities, such as vehicle crashes and blizzard rescues.

MAINTENANCE

Climate change can affect public infrastructure maintenance operations in many ways. Higher temperatures may reduce the number of hours that outdoor maintenance workers can work in hot regions. Thawing permafrost and heavier snow loads may create damages that require repair. Likewise, higher temperatures can also damage infrastructure by exacerbating freeze-thaw cycles or causing road buckling, for instance. Flooding, storm surge, and sea-level inundation may complicate maintenance operations in vulnerable areas. Ice storms that supplant snowstorms in some areas could require different maintenance responses.

TRANSPORTATION INFRASTRUCTURE

Roads and bridges have a variety of vulnerabilities to climate change impacts. As Figures 7.3 and 7.4 illustrate, low-lying infrastructure may be inundated by sea-level rise or storm surge. More frequent or larger floods can damage transportation infrastructure. Bridges designed to specific flood level or sea-level standards may become prematurely obsolete as a result of climate change impacts.

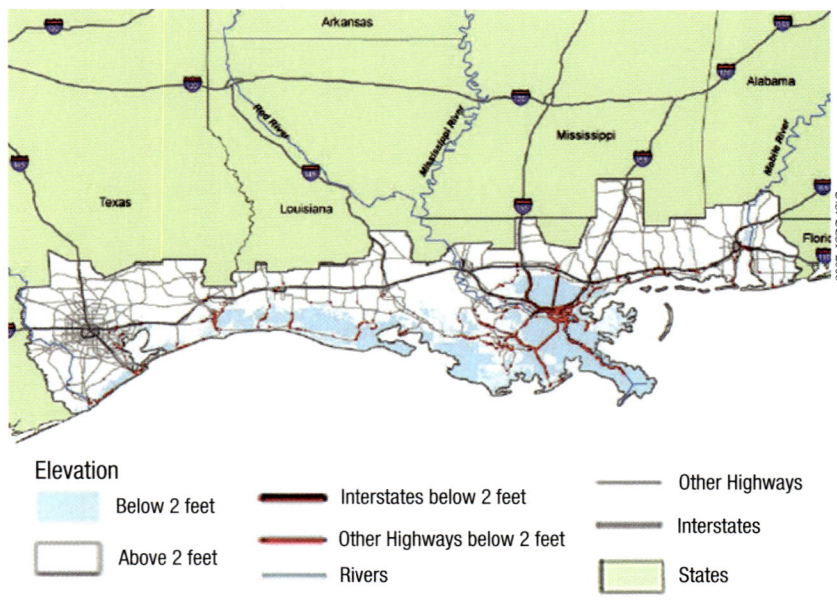

Figure 7.3. Gulf Coast highways at risk from a relative sea-level rise of 61 centimeters (about two feet)

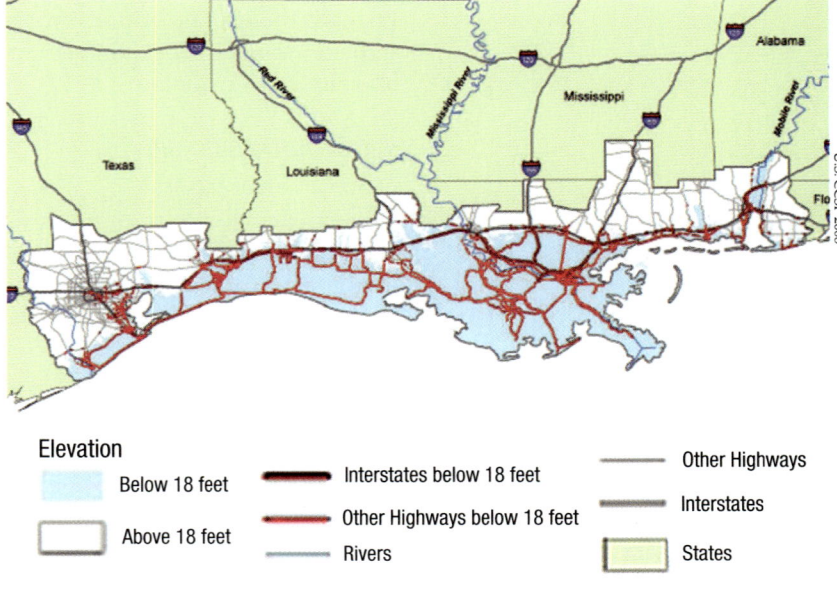

Figure 7.4. Gulf Coast highways at risk from storm surge at elevations currently below 5.5 meters (18 feet)

Sea-level rise has the potential to inundate a significant amount of transportation infrastructure, including roads, bridges, and airports. Ports may be even more vulnerable than roads, bridges, and airports, particularly if their land access routes face inundation. Excessive heat can cause buckling of roadways and railways and can stress bridges.

Figure 7.5. Flooding damage to a road

Not all effects from increasing temperatures will necessarily be negative for transportation, however, particularly in the colder areas of the United States. Decreasing amounts of sea ice could open new routes for transport across polar regions; warmer temperatures may reduce the amount of snow and ice in winter; and less cold weather may result in an extension of the time available for transportation-related construction and maintenance operations.

Communities should be aware of the potential impacts of climate change on roads, streets, and bridges in order to incorporate these design constraints into infrastructure location decisions, bid specifications, design standards, and similar decisions.

WATER INFRASTRUCTURE
Stormwater Management

Climate change will significantly affect stormwater management. Many areas of the United States are projected to experience rainfall and snowfall events of increased intensity, leading to additional runoff. Rapidly melting heavy snowfalls or snowpack due to warmer temperatures can result in erosion, flooding, and water quality impacts from spikes in runoff and reduced infiltration. Landslides and mudslides can result from increased stormwater runoff. In areas where groundwater recharge is necessary to maintain water supplies, lower infiltration may create water supply problems. Additional runoff can also result in additional or more severe flood events.

In some ways, this circumstance mimics the effects of urbanization on stormwater. The impervious surfaces associated with urban development have a two-pronged effect on streams and other natural stormwater systems. First, runoff occurs in greater volumes over shorter durations, creating a higher "spike" than would happen in a natural system, resulting in greater flooding potential as well as negative effects on habitat and natural stream functions. Second, infiltration of precipitation into the ground is reduced by impervious surfaces.

In designing public stormwater management systems and infrastructure, communities should consider the impacts of climate-exacerbated runoff. Green infrastructure approaches that utilize natural systems and regional approaches to stormwater management (as opposed to site-by-site solutions) can be beneficial. Implementation of green infrastructure or regional management solutions may be able to be funded through a stormwater fee-in-lieu regulatory provision that exacts the funds needed to construct on-site stormwater measures and aggregates them into a pool of funds to provide more regional solutions. Funding for such systems may also be available from programs implementing federal water-quality regulations.

An important resource for stormwater management is a community's "green infrastructure"—that is, its network of open spaces and natural areas, such as greenways, wetlands, parks, conservation and preservation areas, and flood-prone areas. Often used to manage stormwater by controlling flooding and improving water quality, green infrastructure also maintains wildlife habitat and travel corridors, provides recreational opportunities, and preserves critical vegetation and open space. Green infrastructure also serves to help mitigate climate change by preserving native vegetation that sequesters carbon dioxide and reducing the urban heat island effect by maintaining tree canopies and streams and other water bodies, resulting in cooler temperatures.

Figure 7.6. Stormwater runoff

Water Supply and Treatment

Adapting to climate change impacts that affect water supply and treatment could be one of the more significant challenges many communities face in the coming decades. Regional impacts that include drought, evaporation, saltwater intrusion, reduced recharge, and flooding have the potential to threaten groundwater and surface water supplies. As supply options become more limited, treatment challenges can occur; for example, more polluted water sources or saltwater sources may need to be pressed into service. Higher temperatures may result in algal and microbe growth that create other treatment challenges. Additionally, water treatment plants, transmission lines, pump stations, and other infrastructure may be located in areas vulnerable to flooding, temporary or permanent inundation, or other climate change–related risks.

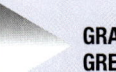

GRAY STREETS TO GREEN STREETS

Because streets constitute a large physical area of public ownership or management, it is important to protect this public investment from climate change impacts to the greatest extent possible. Increased intensity of storm events could create a variety of problems for street infrastructure, including flooding resulting from the overburdening of related stormwater drainage facilities. Street design that incorporates natural drainage techniques—such as swales, rain gardens, and other design elements that enable natural infiltration of stormwater—may offer more stormwater management benefits than will a more structure-oriented approach involving underground drainage lines and catch basins. "Green street" infrastructure should be evaluated as a potential design option in helping a community to adapt to climate change.

Careful assessment of the vulnerability of a community's water supply and treatment is essential to determine future capacity, not only for growth and development but simply for meeting the needs of existing residents, industries, and agriculture. Constraints in water supply and treatment options may result in significant consequences, including shortfalls or limits on future growth. Communities that want to grow or even just maintain their current populations must secure stable future water supplies, a task that may be made more difficult by climate change challenges.

There are a variety of ways of securing stable future water supplies, all of which suffer from the inherent constraint that freshwater supply is ultimately finite. This constraint is obviously less evident in places where water supplies regenerate through abundant rainfall or that have sufficient groundwater storage. However, in areas where there is an imbalance between supply and demand, even limited solutions must be pursued. Examples of efforts to secure stable water supplies include:

- *Increasing access to sources.* The State of Georgia is engaged in a series of lawsuits with neighboring states (Florida and Tennessee) either to withdraw more water than currently allowed under their compacts or to obtain access to new sources—such as the Tennessee River, which Georgia wants to access by redrawing its boundary with Tennessee. On a similar scale, Las Vegas is constructing a $2 billion pipeline to eastern Nevada to access groundwater sources in the Snake Valley. Interbasin transfers across watersheds, state boundaries, and even national boundaries are being considered (see below).

- *Underground storage.* Areas that have enough water at some times and drought at others are considering underground storage of "excess" water in aquifers. Such storage avoids the problem of evaporation from surface storage as well as some of the treatment problems associated with an overabundance of organic matter that can occur in some surface reservoirs. Areas as diverse in climate as Greenville, North Carolina, and Hays, Kansas, are exploring this option. Las Vegas will store its pumped water from the Snake Valley in an aquifer. The underground storage option is not available to all communities because of geologic constraints, and it has its own risks, including pollution of either the "native" water in the aquifer or the stored water.

- *Diversion from other uses.* Figure 7.7 illustrates how freshwater is used in the United States. It shows the potential that diversion has to meet potable water needs, provided that diversion can occur without unacceptable environmental and economic consequences. Some experts believe that inefficient use of subsidized water for agriculture in the Southwest indicates the potential to provide a new "source" of potable water with minimal consequences to agriculture, provided more water-efficient farming practices are established.

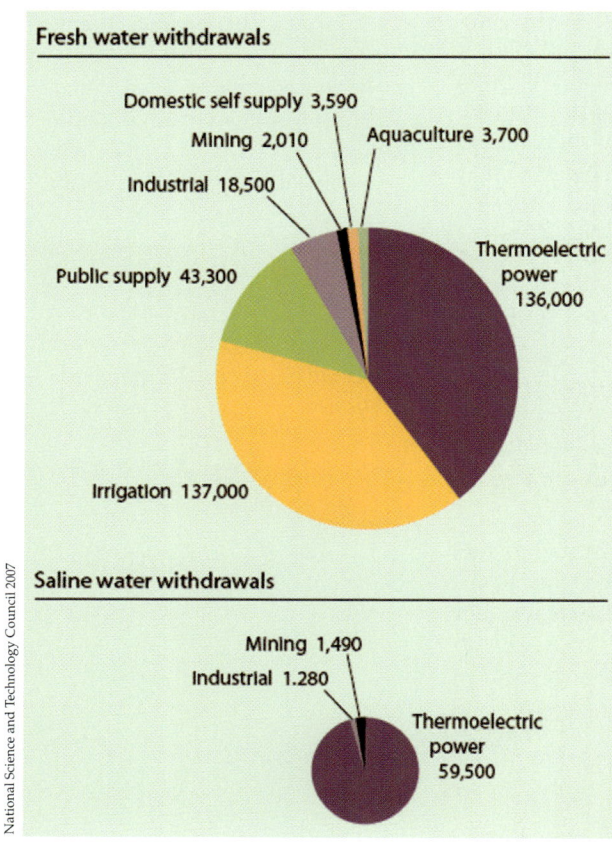

Figure 7.7. Total U.S. water withdrawals, 2000 (million gallons/day)

- *Intrabasin transfers.* Some areas, particularly western states, have experience in negotiating allocations of available water resources; a prime example of this is the recent agreement among the seven Colorado River Basin states (Arizona, California, Colorado, Nevada, New Mexico, Utah, and Wyoming) about the use of the Colorado River water resource. Agreements to share water resources equitably among jurisdictions and use sectors (agriculture, potable, industrial, etc.) may offer solutions to supply problems in some areas.

- *Conservation.* Many communities have experience with voluntary and mandatory conservation efforts during times of drought. In most cases, such efforts have lasted only for the duration of the particular drought, but some communities are making them permanent. Las Vegas, for instance, pays residential property owners to convert their landscaping from turf to xeriscape. Programs to replace older residential toilets with low-flow toilets are being implemented in water-poor areas such as Albuquerque, New Mexico. The use of water pricing schedules, which increase per unit costs for higher-volume users, is another way to encourage the use of conservation measures.

- *Regional interconnectivity.* Establishing connections among separately managed water systems allows water supplies to be shared. With regard to climate change, such interconnectivity is particularly important if it increases the diversity of available water sources, such as a surface-water supply system and a groundwater supply system, since one source may be less affected by drought than another. In some cases, local differences in rainfall or storage capacity are significant enough for interconnectivity to create a desirable "hedge." In others, a significant supply source available to one system may allow multiple systems to avoid future supply problems. Needless to say, interconnectivity is desirable even absent climate change circumstances, in order to avoid problems created by an emergency treatment-plant shutdown, for instance.

- *New technologies.* Using gray water and treated wastewater for irrigation purposes, landscaping for minimal water requirements, using water-efficient appliances and cisterns, implementing industrial water reuse and recycling, and adopting innovative agricultural practices are some of the techniques being used and further explored to make more efficient use of water resources.

Ascertaining a community's vulnerability to climate change impacts affecting water supplies is a complex undertaking. Historic and current information, including forecasts, about rainfall and drought is available through sources

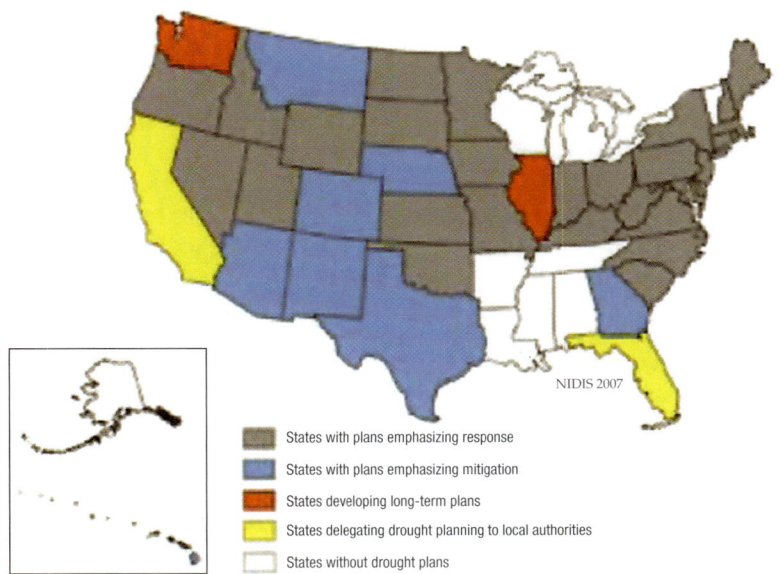

Figure 7.8. State drought plans

such as the U.S. Drought Monitor. Regional climate centers and state climatologists can be contacted for information. Many states have drought plans (see Figure 7.8), which can be accessed. The National Integrated Drought Information System (NIDIS) was initiated in 2006 as a five-year project to create engagement and awareness opportunities and provide information portals that can be customized to user needs (see www.drought.gov).

Regardless of available information, assessing water supply vulnerabilities beyond a general level may require the services of a professional water resources manager, engineer, or hydrologist who is able to account for climate change impacts as well as local circumstances. Regional water use conditions vary significantly, with western states having a higher per capita use than the national average of 179 gallons per day (EPA 2006, 2). Irrigation practices for landscaping and agriculture and the water requirements of industrial users may create significant variations in per capita water use at the local level.

The EPA has issued a number of practical suggestions that should be part of the assessment of water supply availability and should guide responses (EPA 2006). These suggestions are summarized here:

Development Patterns: Large residential lots tend to increase water demand. Various local studies indicate that larger lots create 20 to 60 percent more water demand than smaller ones, largely due to differences in lawn-care requirements. In addition to the pure demand costs, the service-cost differential between large and small lot development is significant. Low density also means more water wasted through leakage: all water systems experience leakage, and the longer service lines required to serve low-density development increase the potential for leakage, increasing the cost of service by reducing the efficiency of delivery. Finally, development on the fringe of urban areas diverts water system resources away from maintenance of existing lines, resulting in continued operation of older, less efficient, or obsolete water infrastructure. The EPA therefore recommends that communities concerned about water resources consider more compact, smart growth development patterns.

Legal Considerations: Public utilities are typically governed by the "duty to serve" provisions of public utility law, which require providing service to any and all customers within a utility's service area. The EPA notes that "the duty to serve can, and at times does, conflict with a utility's or a community's efforts to control water costs and ensure adequate quantities for existing customers . . . sometimes undermining other community goals, such as orderly growth and long-term, stable water provision" (17). Essentially, this requirement places the decision making about service timing and extent in the hands of property owners and developers, rather than local governments and utility system managers. California, Idaho, and Arizona have enacted laws that subordinate duty to serve to comprehensive growth management considerations. As climate change creates new stresses on water supplies, communities should link water supply planning to their comprehensive growth management programs and plans and should encourage growth management considerations to become part of their state's overall water management strategy and legislation.

Financing: One technique that can be used to create an incentive for more compact development is the use of private activity bonding as "a cost-effective way of financing needed water system replacement or upgrades that will support infill development and relieve growth pressures outside the existing system" (19). Another technique involves the establishment of service availability fees (also known as impact fees or facility charges) to cover the marginal cost of new development to the overall water system.

Operations and Pricing Policies: Public utilities should consider a "fix it first" policy that emphasizes maintenance of existing infrastructure rather than expansion into unserved areas, since credit-rating agencies reward utility systems for effective asset management programs, resulting in lower cost for bonded improvements. Additionally, setting rates that fully recoup all system costs is recommended in order to discourage water consumption. In addition, pricing strategies should be considered. Conservation pricing increases water rates for higher-volume users; in some cases, conservation pricing is seasonal or drought related in order to reflect short-term supply problems, but in other cases it is used year-round. Zone pricing establishes rates that reflect service cost differentials for situations such as dispersed development, development at higher altitudes, or large-lot development, all of which have relatively higher service costs.

Finally, the physical vulnerability of water treatment infrastructure needs to be considered in both capital and emergency planning situations. Climate change has the potential to create flooding and inundation problems that affect operations and the viability of water treatment systems on both a temporary and permanent basis. For example, storm surge or floods may interrupt operations, while permanent inundation due to sea-level rise may render some facilities no longer usable. Communities should assess the risks that climate change poses to these facilities and incorporate that information in the planning process.

Wastewater Treatment

Sewage collection and treatment systems will encounter many of the same physical threats that water systems will experience. Sewage treatment plants, collection lines, pump stations, and other infrastructure may be located in areas vulnerable to flooding, temporary or permanent inundation, or other climate change–related risks. Low water flow in discharge areas during droughts may require changes to treatment programs to avoid environmental damage.

In addition to system vulnerabilities, climate change adaptation may require innovations in the use of treated wastewater, such as for irrigation purposes. Such adaptation measures may require new treatment protocols and will certainly require new infrastructure.

Joint stormwater and sewage treatment systems (known as combined wastewater systems) create special health and environmental problems in flood conditions. These systems are designed to divert wastewater that exceeds plant treatment capacities back into natural systems during floods. This creates the potential for contamination of downstream water supplies with pollutants and waterborne diseases. Potential contaminants include *Cryptosporidium* (a protozoa that causes intestinal illnesses), *E. coli* (a bacterium that causes intestinal illnesses), pesticides, and petroleum products. Figure 7.9 illustrates the general location of these systems throughout the United States. Even sewage treatment facilities that are not combined with stormwater systems can become inundated during flood conditions or by storm surge, creating similar problems.

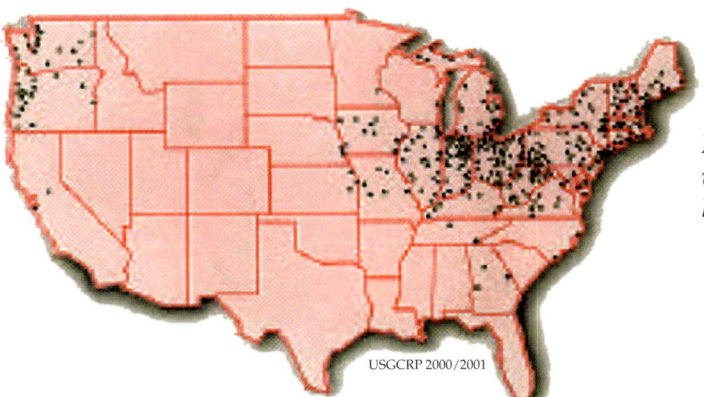

Figure 7.9. Combined wastewater system locations

ANN ARBOR: A SELF-SUSTAINING ENERGY FUND FOR ENERGY EFFICIENCY IN MUNICIPAL FACILITIES

Population: 114,386

The Municipal Energy Fund was established by the city in 1998 to be a self-sustaining source of funds for investment in energy-efficient retrofits at city facilities. The principle is smart and simple: for each energy-efficiency investment project, the fund provides the up-front costs, which motivates facility managers to invest in energy efficiency. These investments are recouped by repaying the energy fund 80 percent of the cost savings resulting from the improvements for a period of five years. This five-year payment period allows projects that have a shorter payback (three years or less) to help support projects that have longer ones.

According to city government sources, the critical points to ensure the success of the project were to find an initial funding source and to assign a manager to support and coordinate the fund. The $100,000 per year initial funding has proven to be adequate to the size of the city (about 60 facilities with $4.5 million per year in energy costs). The city incurred no additional costs to manage the Municipal Energy Fund. The implementation of the program became one of the duties of the manager of the city's energy programs, an existing staff position.

Communities should assess vulnerabilities to sewage collection and treatment systems that result from climate change. These vulnerabilities include flood risks, collection line infiltration and outfall problems resulting from intense rainfall, temporary inundation occurring from storm surge, and permanent inundation problems from sea-level rise. Planners should also assess the potential for wastewater treatment systems to assist in addressing water supply issues, such as the use of treated wastewater for irrigation purposes.

Energy Issues Associated with Water and Wastewater Treatment

Collecting, storing, and treating water requires energy for pumping and treatment processes. Conservation measures offer an opportunity to reduce not only the amount of water needed but also the energy cost of producing that water; such measures include changes in water use practices and habits, use of low-flow fixtures, regulations, public awareness campaigns, and so on. Improved maintenance techniques that reduce system leakage will improve the efficiency of water delivery and reduce the amount of energy needed to produce the water that actually reaches the end user. On-site generation of power through hydroelectric, solar, and wind sources can be used to reduce the carbon-based energy costs associated with water production. Water reuse and recycling programs can reduce water treatment energy costs. Finally, more energy-efficient pumping and treatment processes can be implemented.

ENERGY INFRASTRUCTURE

Electric Generation, Transmission, and Distribution

Increasing the energy efficiency and share of renewable energy used to generate electricity is a cornerstone of facing energy and climate issues. Understanding the different planning, siting, and operation requirements of different renewable energy and energy-efficiency technologies can help maximize opportunities to deploy these technologies.

Electricity is only valuable when it can cost-effectively get to where it needs to be used; hence, the relationships among generation, transmission, and distribution are critically important. Transmission refers predominantly to concentrated high-capacity and high-voltage conveyance. The high voltage allows electric current to be carried over long distances with minimal "line-loss"—the drop in efficiency from conveyance. However, this requires transformer substations that "step-up" the current for high-voltage transmission and "step-down" the current nearer to where it is "distributed" to its end use.

New large electric generation facilities may require new transmission and generation capacity, and renewable energy is no different. Two measures are particularly important in assessing existing transmission and distribution capacity when planning new renewable energy facilities. First is the total carrying capacity of the transmission and distribution grids involved. Electricity "likes" to have plenty of room to move through, meaning that when the average electric load of a system nears the capacity of the system, efficiency decreases. Efficiency starts to drop at approximately 80 percent of total capacity.

In practice, peak-load capacity is frequently a more significant factor. Electricity is subject to "bottlenecks" at junctures in transmission and especially in distribution. Once electricity gets beyond substations or transmission bottlenecks, distribution may occur over longer distances as well, but with an efficiency penalty that may be up to 0.5 percent per mile. From an electric grid perspective, compact development in proximity to substations is a desirable goal.

Or, one can circumvent the transmission substation, step-up/step-down, potential line-loss approach altogether and locate generation facilities relatively close to end uses. This is one of the principal benefits of decentralized or "distributed" (in a different but related sense of the word) power generation, which many renewable energy technologies can provide.

Renewable energy and energy efficiency are paramount goals for both large and small facilities and for centralized and decentralized approaches. Ways that utilities can promote efficiency and conservation include assisting residential and business customers with energy audits, offering rebates for Energy Star appliances, instituting conservation awareness programs, subsidizing the purchase of compact fluorescent lightbulbs and high-efficiency heating and cooling systems, and introducing remote peak-load management of certain appliances and HVAC systems (such as refrigerators, hot water heaters, air conditioners, and electric heat pumps), allowing these appliances and systems to be turned off for short periods of time to moderate peak-period energy consumption.

District Heating and Cooling

Increasing use of renewable energy and energy efficiency may also involve the installation or upgrading of pipe networks to carry thermal energy via water or special high-thermal-capacity fluids. The primary reason is that the operation of heating and cooling facilities can be carried out much more efficiently in a moderately centralized system than in a configuration of individual heating units and air conditioners for individual buildings. The efficiency gains are on the order of 10 to 30 percent.

The siting of centralized or "district" heating and cooling facilities and accompanying pipes is shaped mostly by the configuration of existing buildings and where pipe right-of-way is available. Existing pipe right-of-way, especially in larger, urban settings, is a critical factor in the viability of new district heating-cooling systems, as excavation to install new pipes can be costly in retrofitting existing developments. Large development and redevelopment projects, however, may offer multiple opportunities to install pipes at little to no additional excavation costs.

The source of the heating and cooling energy is the other critical factor. Combined Heat and Power (CHP), where the thermal heat is a by-product of producing electricity, is a largely untapped source. Biomass and geothermal sources are the most common. For biomass-fueled plants, the main issues are simply finding a suitably large site that allows relatively easy delivery of biomass for the plant. Biomass facilities do have emissions issues that need to be addressed. Though most existing state regulations address these issues, additional local issues may arise, primarily regarding the location and height of the exhaust stack.

TOOLS FOR GUIDING INFRASTRUCTURE DECISIONS

Visioning and Goal Setting	Engage transportation, engineering, and public works staff and establish asset management goals that incorporate costs and mitigate risks of climate change
Plan Making	Inventory local energy resources and efficiency opportunities and integrate solutions to climate hazards to infrastructure into the comprehensive plan
Standards, Policies, and Incentives	Establish policies and incentives that facilitate energy and water efficiency
Development Work	Encourage development in areas served by existing infrastructure and in low climate-risk areas
Public Investment	Prioritize public investments in infrastructure that avoid areas subject to flooding and other climate-related risks

CHAPTER 8

Transportation

 This chapter discusses strategies for promoting energy efficiency, renewable energy, and greenhouse gas reduction in the transportation sector (adaptation issues facing transportation infrastructure were addressed in Chapter 7). Studies suggest that no single measure is enough by itself to achieve energy and GHG savings commensurate with national goals (Cambridge Systematics 2009). Rather, transportation requires the application of a suite of integrated and complementary measures. These measures fall into four main categories:

- Managing Travel Demand
- Maximizing System Efficiency
- Shifting Travel to More Energy- and Carbon-Efficient Modes
- Promoting Efficient Vehicles and Low-Carbon Fuels

MANAGING TRAVEL DEMAND: THE TRIP NOT TAKEN

The most energy- and carbon-efficient trip of all is the trip not taken. Transportation is typically not an end unto itself but a means by which we meet the needs of our daily lives. While some trips may be "just for the fun of it," most trips are functional—there is a need to be met or a purpose to be fulfilled. So the first principle of energy efficiency for transportation is to provide options to not travel—thereby reducing the number of trips and miles traveled.

For work trips, providing enhanced options for telecommuting—via both policies to promote it and technology to make it convenient and viable—is an important strategy. The number of Americans who worked remotely at least one day per month for their employer ("employee telecommuters") increased from approximately 12.4 million in 2006 to 17.2 million in 2008, and this trend is expected to accelerate (World at Work 2009). For many, telecommuting simply requires setting up appropriate connections between a home computer and an office network and having the freedom to work at home on a regular or occasional basis. Others, however, may not readily have the equipment, space, desire, or option to work at home. Programs and incentives to enhance a home-based work environment or provide amenities that expand options to work at other private and public places closer to homes—such as public libraries, coffee shops, or even designated telecommute centers—can be much more cost-effective and energy-efficient options than conventional travel.

For shopping trips, which account for approximately 20 percent of all vehicle trips (Hu and Reuscher 2004), enhancing and promoting options for remote shopping and home delivery can reduce and obviate individual trips. While online shopping is now commonplace for many people, consumer surveys and industry observers have indicated that significant potential remains to improve service and expand participation (Pew Internet 2008). Most major supermarket chains, for example, are expanding online ordering and home delivery services. Other local businesses may not offer their own delivery services, but third-party companies that deliver a variety of retail goods are also becoming more common—and such trips can be more energy efficient when businesses are able to chain trips together. Market research suggests that online ordering from local businesses has the potential to become as prevalent as shopping from national online retailers. Additional incentives for private businesses to expand delivery options, including starting a general-purpose delivery service, may be cost-effective ways to enhance transportation efficiency, though this depends on many factors, especially the current pattern and nature of trips and the land-use context. Local travel behavior surveys can be helpful in assessing the potential efficiency gains of delivery service.

Some trips—school-related, social, recreational—may not be readily replaced by other options. However, on occasion, the purposes of these trips may be achieved remotely. Many universities have increasingly made course materials available online or offer entire distance-learning curriculums. The goal is not to eliminate all trips and the social interaction that comes with them but to increase the flexibility to not travel, unless travel is specifically desired or necessary. This enhanced flexibility also may prove to be valuable in the future, when travel routes and normal gathering activities may be disrupted more often by strong storms, public health emergencies, and other impacts that climate change may exacerbate. Community resilience to the effects of travel disruptions must be bolstered in the face of climate change.

Pricing

Studies suggest that travel demand can be strongly influenced by different kinds of pricing mechanisms and other incentives. In fact, according to one study, pricing may be the most effective single tool for curbing travel demand, with potential reductions in VMT on the order of 10 to 15 percent (Cambridge Systematics 2009). Many elements of automobile transportation—including insurance, parking, and road use—are priced at a fixed rate or not at all, regardless of time of day or distance traveled—the equivalent of "all-you-can-eat" pricing. Studies of pricing mechanisms—such as pay-as-you-drive insurance, roadway tolling, and variable parking rates—show that drivers would adjust their amount of vehicle travel in response to a clear price signal (Bordoff and Noel 2008; San Francisco County Transportation Authority 2008).

Transportation Demand Management

An employment-oriented transportation demand management (TDM) program might include outreach to major employers to discuss adjustments in shift schedules to allow off-peak commuting, the provision of support for vanpools and carpooling, the accommodation of on-site or nearby transit access, the creation or enhancement of telecommuting programs, and similar measures either to distribute trips more evenly during the workday or to reduce VMT. Transportation demand managers also should attempt to become involved in industrial recruitment efforts in order to provide critical and comprehensive transportation support for potential new industries.

MAXIMIZING SYSTEM EFFICIENCY

After minimizing travel demand, maximizing the efficiency of existing transportation systems and travel patterns is the next most immediate measure for reducing transportation-related energy use and GHG emissions. For a given volume of cars, buses, trains, and people moving through a transportation network over a given time period, the essential goal for system-level efficiency is distributing that volume to allow the smoothest flow possible. Distribution is managed in four key ways:

- Peak Pricing and Incentives
- Street Connectivity
- Transit-Oriented Development
- Congestion Management Strategies

Peak Pricing and Incentives

In addition to its travel demand effects, pricing can be an important tool for distributing traffic flow. Tolls and fees for using certain roads or entering certain areas can help divert traffic to other routes or modes of travel, thereby promoting a more efficient flow of travel throughout the system.

Efficiency in the transportation system can be improved by other behavioral incentives as well—such as by employers offering flexible work hours to their employees, which can help reduce traffic congestion during peak rush hours. The adoption of transit incentive programs and the provision of bicycle parking and facilities (such as lockers and showers) are other ways to encourage travel in nonautomotive modes, thereby distributing trips and mitigating congestion.

Street Connectivity

Street connectivity is another important parameter that affects all modes of travel. Street connectivity creates a range of opportunities to reduce VMT. Connected streets can result in shorter trips, which not only reduce driving distances but also make it more likely for people to use bicycle and pedestrian travel modes to get to their destinations.

Planners can work to improve the connectivity of local streets in a variety of ways. Subdivision regulations and technical standards and specifications manuals can contain requirements that support street connectivity, such as short block lengths, grid street-pattern requirements, maximum cul-de-sac lengths, and similar regulatory standards. Capital improvement programs can include projects that connect existing streets.

Transit-Oriented Development

In communities with light rail or extensive bus transit service, TODs are often used to create nodes of intense development and redevelopment centered around transit stations. The transportation efficiencies of siting dense residential development and workplaces within walking distance of transit stations are obvious—fewer commuter trips are made by personal vehicle. If shopping and entertainment uses are included as part of TOD design, personal vehicle trips can be further reduced.

Figure 8.1. A Portland, Oregon, streetcar

Central business districts and downtown areas should be designed to accommodate mixed use development and to cultivate an urban form that incorporates pedestrian traffic as a priority mode of travel (e.g., by including wide sidewalks, attractive streetscaping, sidewalk-facing facade orientation, interesting ground-level shops and restaurants, and on-street parking separating pedestrians from vehicular traffic). Where feasible, transit centers should be located in downtown settings to promote transit access to and from downtown and to provide transit service to downtown businesses and offices.

Suburban locations can offer opportunities to reduce VMT as well. Concentrated areas of retail development and denser neighborhoods can be efficiently served by transit. Greenways, bicycle lanes, sidewalks, and other pedestrian facilities can be installed to encourage multiple modes of travel while also serving commuter needs.

Congestion Management Strategies

Managing congestion offers opportunities for reducing CO_2 emissions. Causes of congestion are varied and include capacity-related bottlenecks, inclement weather, traffic incidents (mainly crashes), poor signalization, and road construction.

The Texas Transportation Institute (http://tti.tamu.edu) has identified a number of congestion management programs that communities can adopt:

- Access management programs that minimize or combine driveway cuts, increase spacing between intersections, provide medians and turn lanes, and provide acceleration/deceleration lanes.

- Traffic signal coordination programs that synchronize timing of traffic signals for optimum flow.

- Incident management programs that provide more rapid responses to accidents and weather events.

- Programs to increase capacity, including roadbuilding and time-of-day travel lane shifting that supports the dominant travel-flow pattern.

- Programs to increase the availability and amount of public transportation.

- Programs to relieve bottlenecks, dysfunctional intersections, and merge points.

- Travel option programs, such as instituting HOV lanes, which reward carpooling and ridesharing.

Another source for congestion management strategies is the MARC Enhanced Congestion Management System developed for the Mid-America Regional Council, which can be accessed at www.marc.org/transportation/cms/toolbox.pdf.

SHIFTING TRAVEL TO MORE EFFICIENT MODES

Another important energy- and greenhouse gas–saving strategy is to shift passenger or freight travel to less energy- and carbon-intensive modes. For passenger trips, that typically means shifting automobile trips toward transit, biking, and walking trips or to multimodal trips that may involve car, bike, transit, and walk segments. For freight, it means shifting at least some portion of long-haul trips from truck to rail or ship travel.

Transit

For larger communities and more compact smaller ones, transit is an extremely significant tool for reducing VMT on a per

PARKING

A significant portion of urban land is devoted to parking. Negative climate change impacts include higher urban temperatures from the heat island effect and stormwater management problems from increased rainfall falling on impervious areas in urban areas. Reducing the parking "footprint" is an important adaptation response.

Many parking regulations, particularly those that calculate required parking on the basis of the needs of each individual use without regard for time-of-use patterns, result in the provision of too much parking. For example, a development containing office uses and a movie theater has few overlapping hours of parking for the two uses; the parking provided for the office uses is largely empty when the theater parking is at peak demand and vice versa. A shared parking ordinance would recognize this use pattern in establishing the required number of spaces.

Figure 8.2. A mixed use parking structure in Kirkland, Washington

Reducing the footprint of parking on a site reduces the impervious surface and can result in cooler development. Minimizing the number of parking spaces to the level necessary to handle actual parking demand through shared parking requirements, maximum parking requirements, and similar standards creates a balance between the amount of parking provided and the need for such parking. Another technique is to allow on-street parking to count toward meeting minimum off-street parking requirements.

Some communities encourage a reduced parking footprint by promoting the use of structured parking. This is often done by exempting structured parking from maximum parking requirements or by providing incentives for parking under structures (often for structures that must be elevated to meet flood or storm surge protection requirements). Allowing structured parking to utilize valet or mechanical or robotic parking methods can reduce space requirements since vehicles can be parked closer together. A number of communities simply require all parking in certain zoning districts to be in structures. Downtown or central business district zoning regulations in many communities contain no minimum parking requirements. In parts of San Francisco, certain uses have strict maximum parking requirements to discourage commuter driving and to promote transit use.

capita basis. Since transit requires a certain level of residential density to be effective (the minimum density threshold necessary to support transit is typically considered to be eight dwelling units per acre), more dispersed communities may not be able to effectively implement a transit program except in limited circumstances, such as a downtown shuttle or carpooling and park-and-ride programs serving major places of employment.

Effective transit programs require a comprehensive management approach. In addition to providing transit vehicles, route planning, functional maintenance of vehicles and stations, and employees such as operators and maintenance staff, good transit programs address smaller customer-service issues as well. For example, a major impediment to transit use can be removed or reduced by offering backup taxi or van service to transit riders in emergency situations, such as an illness or a pressing child care issue; such services limit the vulnerability some people may feel when relying on transit as their primary transportation.

TravelMatters.org, a project the Center for Neighborhood Technology prepared for the Transportation Research Board, documents examples of good transit service, marketing, and operations programs. It also provides an assessment of the importance of good management in overall transit-program effectiveness:

> In many places, people drive not because they want to, but because there are few practical alternatives. Where transit options do exist, poor transit service, management and marketing often fail to attract potential riders.... In the short-term, there are many low-cost actions open to transit agencies to make the transit experience more pleasant for the public, whether this means maintaining the interior and exterior cleanliness of a vehicle, customer service training for personnel, or providing efficient and comfortable means of access and egress to vehicles at transit stops.

There are many cases where municipalities or districts have used innovative methods to implement transit programs, including:

- The Minneapolis/St. Paul region's Metro Transit is heavily promoting a program to encourage less driving and more use of transit that is linked to the Minnesota Energy Challenge, a comprehensive carbon footprint reduction program. For more information, see www.mnenergychallenge.org.

- Portland, Oregon, established a new streetcar line in 2001; the $89 million infrastructure project has helped trigger more than $2.5 billion in new private-sector investment, including the successful Pearl District redevelopment.

- On high-ozone days, some communities like Asheville, North Carolina, offer fare-free rides as a means of combating air pollution while raising awareness about the link between driving and air pollution. Other communities offer fare-free zones, as Denver does in its downtown, to encourage people to ride transit in typically congested locations.

- Some communities utilize a range of development exactions to promote transit use, such as requiring developers to install on-site or off-site transit facilities, provide transit access to the development site, participate in specialized transit services (such as a trolley service within a large shopping area), or purchase transit passes for employees or residents of the development.

PORTLAND, OREGON: A LEADER IN SUSTAINABLE TRANSPORTATION SYSTEMS
Population: 550,396

The case of Portland, Oregon, demonstrates the important role of planning in providing the foundation for a successful multimodal transportation system that has important climate change and energy efficiency implications. Since the 1980s, the city has been investing in pedestrian, bicycle, and public transportation improvements to increase transportation options for its citizens. In recent testimony before the U.S. Congress, Catherine Ciarlo, Portland's acting transportation policy director, testified that between 1996 and 2006 transit ridership in the Portland Metro region increased by 46 percent, while daily vehicle miles traveled per capita declined by 8 percent (Ciarlo 2009). Portland has consistently chosen to expand transportation alternatives and, as the statistics show, has been successful in shifting passenger miles from vehicles to other, less carbon-intensive modes.

The Portland/Multnomah Climate Action Plan, first released in 2001 and last updated in 2009, draws upon Portland's transportation successes and connects transportation planning to emissions-reduction goals. The plan is the product of a collaboration of city and county agencies, business, nonprofit organizations, and members of the community. According to the plan, by 2007, the Portland region had reduced carbon emissions 1 percent below 1990 levels, despite rapid population growth and an upward national trend in carbon emissions. Portland is one of the few U.S. cities to successfully quantify and track progress toward emissions-reduction goals.

U.S. Environmental Protection Agency statistics show that transportation is the fastest-growing source of U.S. greenhouse gas emissions. Widespread urban sprawl and America's heavy reliance on the automobile are major contributors to this trend. GHGs from transportation account for 47 percent of the net increase in total U.S. emissions since 1990 (EPA 2009d). In Portland, by contrast, GHGs from transportation decreased slightly from 1990 to 2008. Nonetheless, even in Portland, transportation accounts for 40 percent of all GHG emissions and presents substantial opportunities to reduce emissions.

The Portland/Multnomah Climate Action Plan sets aggressive goals targeting further reductions in emissions from transportation. By promoting compact, mixed use development patterns, the city hopes to create neighborhoods where 90 percent of residents can easily walk or bicycle to meet all basic daily needs. Continued expansion and upgrades of the existing alternative transportation network will be complemented by existing state and local policies that promote compact development and encourage alternatives to vehicles.

Pedestrian and Bicycle Infrastructure

In the early 1970s, the decision to reclaim Harbor Drive from automobiles was one of the city's earliest attempts to shift passenger miles from cars to alternative transportation modes. The six-lane road running along the downtown waterfront was removed and replaced by Waterfront Park, with extensive provisions for pedestrians and bicycles.

Figure 8.3. (Above) Bicyclists on the Hawthorne Bridge, Portland, Oregon

Figure 8.4. Portland's light rail

The city has worked to improve and upgrade facilities to encourage walking and biking as alternatives to automobile transportation. Portland now has a bicycle commuter rate of 8 to 12 percent—a higher percentage than that of any other major U.S. city and eight times the national average (Ciarlo 2009).

Public Transportation Infrastructure

In 1986, Portland opened the first line of its light-rail system, the Metropolitan Area Express (MAX). Construction of the system was chosen in lieu of a proposed infrastructure expansion for vehicles, the Mount Hood Highway. MAX is just one component of the TriMet transportation system that services the Portland metro area with light rail, bus lines, and commuter rail. TriMet carries more people than any other U.S. transit system its size. Weekly ridership on buses and MAX has increased for 20 consecutive years. Since its inception, the MAX system has been a resounding success, receiving both national and international accolades.

(continued on page 94)

(continued from page 93)

The Portland streetcar system was launched in 2001. It connects major destinations in the central city and links to regional light rail and bus systems. The introduction of the streetcar allowed for higher-density development and lower parking ratios in the areas served by it. The project was made possible through a combination of private funding, revenue from parking fees, creation of a local improvement district, and tax increment financing (TIF).

The neighborhoods that have developed around the streetcar and light rail are striking for their walkability. In addition to environmental and health benefits, the advent of alternative transportation has revitalized neighborhoods and had positive economic benefits by increasing property values and retail revenues around transit-oriented developments.

Oregon Bike Bill

The state has played an invaluable role in strengthening Portland's transportation systems. The 1971 Oregon "bike bill," ORS 366.514, requires the inclusion of facilities for pedestrians and bicyclists wherever a road, street, or highway is built or rebuilt by a municipality, county, or the state Department of Transportation.

TOD Tax Abatement

The Transit Oriented Development (TOD) Property Tax Abatement was adopted by the Portland City Council to encourage and incentivize compact development near transit. Projects that are located within a quarter mile of a light-rail station or selected corridors qualify for a tax exemption for a ten-year period, based on the value of the project improvement (Portland Development Commission 2006).

Further progress can still be made in Portland, but the city has distinguished itself with its commitment to improving and diversifying transportation options and has developed a considerable reputation for innovation in comprehensive land use and transportation planning. Coordinated planning and policy, strong local leadership, and public-private partnerships have all contributed to implementing the city's vision of a sustainable multimodal transportation system. ◄

Biking

Bicycling can be an important VMT reduction tool. However, there are some impediments to bicycle travel:

- *Travel distance.* Sprawling development patterns result in longer travel distances for all transportation modes, including the bicycle.
- *Travel safety.* Most U.S. streets and roads have been designed largely for the exclusive use of the automobile, creating safety problems that bicyclists must overcome through route selection and other exercises of personal diligence.
- *Travel convenience.* Other modes of travel may be more convenient than bicycling due to weather, road congestion, topography, travel distance, or safety issues.

Planners seeking to reduce VMT by promoting increased use of bicycling need to address these three impediments in order to be successful. The use of more compact development patterns, road designs that are able to safely—and pleasantly—accommodate bicycle and pedestrian traffic, and more convenient bicycle route layouts and bike-specific services can encourage more frequent bicycle travel. Since compact development patterns and complete streets are discussed elsewhere in this section, only examples of possible convenience improvements are provided here:

- Creating a mix of on-road and off-road (e.g., greenways, trails, etc.) bicycle routes can provide more choices for convenient travel.
- Providing storage facilities or sturdy bicycle racks allows for bicycles to be safely stored at their riders' destinations. Many communities have bicycle parking requirements that are implemented much like automobile parking standards.
- Starting a community or corporate bicycle fleet–sharing program makes cycling more convenient. Some existing programs utilize membership fees, special rental stations, and trip pricing based on the length of use. Funds received pay for acquiring and maintaining the bicycle fleet.
- Adding bicycle carrying racks to transit vehicles makes it easier to link multiple travel modes.
- Providing emergency taxi or van services to bicycle riders (as suggested for transit users) reduces the vulnerability of cyclists relying on a single travel mode.
- Providing workplace showers and changing facilities improves convenience for bicyclists.

The New York City Department of Transportation has tracked a 75 percent increase in bicycle trip volume in central Manhattan between 2000 and 2006. Between 2007 and 2008, New York saw commuter cycling increase by 35 percent in its central business district (see www.nyc.gov/html/dot/html/bicyclists/bikemain.shtml).

Walking

Pedestrian travel in many areas of the United States is complicated by a lack of sidewalks, poor connectivity of the street network, intersections without pedestrian improvements, and segregation of land uses that results in long travel distances. Planners' efforts to create more

Figure 8.5. New guidance on bicycle crossings, Portland, Oregon

compact development patterns, complete streets, greater street connectivity, and greater diversity of land uses can help address these problems.

Moreover, pedestrian environments need to be safe, functional, and attractive. There are a variety of elements that can improve pedestrian safety and ensure an inviting pedestrian environment. Examples include:

- Accessible sidewalks
- Tree-lined streets
- On-street parking, which provides a buffer between pedestrians and vehicles in transit
- Off-road pedestrian facilities such as greenways and trails
- Traffic-calming practices
- Pedestrian-scale and pedestrian-oriented development, which places parking at the rear of buildings instead of at the front and orients storefront windows and signage to pedestrians

Figure 8.6. Walking and biking to school in Boulder, Colorado

- Pedestrian connections among destinations, such as schools, shopping, restaurants, and other neighborhood-oriented services
- Barrier-free design that ensures access to the pedestrian system by all users, including those with disabilities or special needs.

Multimodal Approaches

Transportation networks that accommodate multiple modes of travel are an important tool in transitioning trips to more GHG-efficient modes. Complete streets, for example, accommodate pedestrians, bicyclists, transit, and automobiles. In order to accommodate these various modes safely, complete streets may include sidewalks, bike lanes, bus lanes, crosswalks, medians, bus pullouts, and sidewalk bulbouts.

In order to facilitate multimodal transportation networks, communities may need to revise existing policies or standards or adopt new policies and standards. The overall transportation network in a community should also be examined to see where improvements are needed and ensure overall connectivity. It may not be practical to make improvements on every street, but careful examination of the overall network will help to identify priority routes and segments so that all modes can travel conveniently throughout a community and a region.

Freight

Different modes of transporting goods have very different energy costs and GHG emissions associated with them. Freight handled by ships and rail is more efficient from an energy standpoint than freight handled by trucks. For example, the carbon footprint of a bottle of French wine shipped across the Atlantic to New York City is 2.93 pounds of carbon, compared with 7.05 pounds of carbon for a bottle of California wine trucked to the same location (Doukoupil 2009); the difference is entirely related to the energy cost of the different shipping modes. Therefore, opportunities to shift freight transport to more GHG-efficient modes can help regions reduce carbon emissions.

PROMOTING EFFICIENT VEHICLES AND LOW-CARBON FUELS

Improving the efficiencies of vehicles and carbon-intensity of fuels is an important way to reduce energy use and carbon emissions. Local communities and planners are generally not in positions to develop more advanced vehicles and fuels, but they can provide incentives and other measures to promote and encourage their use. Important measures include:

- Supportive Infrastructure and Preferential Lanes
- Parking
- Fleet Purchasing and Management

Supportive Infrastructure

Communities can provide and encourage the provision of alternative fueling and recharging stations, which can help promote the use of automobiles that require such services. Designating travel lanes for high-occupancy vehicles, electric vehicles, and vehicles that utilize low-carbon fuels is another strategy that can encourage a transition to more energy- and GHG-efficient vehicles.

Some communities have begun installing prominently-located public charging stations. Providing electricity for a few cars in a few places is one thing, but expanding to thousands of cars will require an assessment of the available capacity to supply additional electricity as well as a method for proper charging and accounting of the electricity consumed. Not surpris-

FORT COLLINS, COLORADO
Population: 129,467

Fort Collins is the fifth-largest city in Colorado and is home to the 25,000-student main campus of Colorado State University. Fort Collins has a diverse economy that includes breweries as well as biotech and clean energy companies. In recent years, various publications have ranked it as one of the best places to live in the United States. Adding to the desirability to live here is the city's strong commitment to making alternative transportation modes, such as walking and biking, safe and easily accessible.

Over the years, Fort Collins has taken a highly proactive stance in encouraging a diverse set of transportation options. Since the mid-1990s, the city has defined a successful transportation system as one that maximizes and balances access, mobility, safety, and emergency response throughout the city, while reducing vehicle miles traveled and dependence upon the private automobile. The Fort Collins government has also recognized the connection between reducing VMT and reducing GHG emissions, as demonstrated in its Comprehensive City Plan (2004), Transportation Master Plan (2004), Bicycle Plan (2008), and Climate Action Plan (2008), among others.

The City Plan established sustainability, fulfillment, fairness, and choices as formal community values. The community vision identified in the City Plan interprets those values as the basis of a mandate to address all modes of transportation and encourage development patterns that are conducive to pedestrian, bicycle, and public-transit travel.

The 2004 Transportation Master Plan reaffirms the city's commitment to providing a multimodal transportation system. The plan recognizes that a rise in vehicle trips has led to an increase in both GHG emissions and severe congestion. To address this issue, Fort Collins is committed to providing a more balanced transportation system. The plan also provides updates to modal plans that have been developed and more clearly defines other transportation ideas, such as the concepts of enhanced travel corridors, context sensitive design, and complete streets.

The 2008 Bicycle Plan updates and expands upon the 1995 Fort Collins Bicycle Program Plan. The goal of the 1995 plan was to make bicycling an easy transportation option. The 2008 plan builds upon that goal through initiatives such as the Safe Routes to School program, which encourages students to walk or ride their bikes to school.

The Climate Action Plan that Fort Collins adopted in December 2008 also addresses the need to provide more transportation options in order to reduce VMT and GHG emissions.

This emphasis on expanding transportation choices has had an impact on the urban form and function of the city. Fort Collins now has 280 miles of bicycle lanes, 30 miles of hard-surfaced, multi-use paths, and many more miles of local-street bicycle routes. Future bike-lane projects will take place with new street construction or reconstruction of existing facilities per the city's Master Street Plan. In the works are cooperative projects with local businesses to install on-street bicycle parking, which helps mitigate sidewalk congestion and encourages better pedestrian flow by eliminating sidewalk obstacles.

Matt Wempe, a transportation planner in the city, reports that the FC Bikes program has enjoyed a tremendous renaissance in the past three years. During the last city budget cycle, the public pressed the city council to reinstate the bicycle coordinator position that had been cut in previous budgets. As Wempe says, "Fort Collins views bicycling as a way of life." FC Bikes hosts an annual Bike Week in June that includes a Summer Bike to Work Day and Bike Prom. It also sponsors a Winter Bike to Work Day, developed a unique education campaign called "Co-Exist," created the Bike Library lending program, and supports the many local bicycling community organizations.

Figure 8.7. *Bike to Work Week in Fort Collins, Colorado*

The biggest measure of success came last year when Fort Collins was designated a gold-level bicycle-friendly community by the League of American Bicyclists. Only nine other communities nationwide are currently designated at that level. Bike to Work Day 2008 garnered 8,000 participants who biked more than 22,000 miles. Several events also exemplify how the bicycle community has grown, including weekly bicycle races, the summer B.I.K.E. Camp for kids, and the Fossil Creek family bike fair. There's also the regular occurrence of having more bikes than cars at any given intersection.

Fort Collins has vigorously pursued grant monies for bicycle improvements from sources such as the federal Congestion Mitigation and Air Quality (CMAQ) and Transportation Enhancements (TE) programs, as well as Great Outdoors Colorado (GOCO) lottery funds. Since 1995, the city has secured more than $20 million in federal grants. The CMAQ funding requires the city to quantify the air-quality benefits of bicycling, and these data are also used to determine progress on the city's Climate Action Plan.

The city's transit system, Transfort, has bike racks on all buses, provides service to CSU students through university-collected fees, and covers 18 daytime routes with 23 buses during peak service. Transfort is currently funded by ridership revenues, the city general fund, and federal grants. Additionally, a cooperative effort among the cities of Loveland and Fort Collins and Larimer County funds FoxTrot, a route linking the two communities.

(continued on page 98)

(continued from page 97)

Another development of note in Fort Collins is the Mason Corridor, a five-mile north-south byway located along the Burlington Northern Santa Fe (BNSF) Railway property just west of College Avenue (U.S. Route 287). Sixty percent of Fort Collins residents work within one mile of the Mason Corridor, making it the main link connecting downtown, the university campus, and the shopping area. The corridor includes a new bike and pedestrian trail as well as a planned Bus Rapid Transit (BRT) system in a fixed guideway for the majority of the corridor. The BRT service will operate nearly twice as fast as auto travel along College Avenue, with service as frequent as every 10 minutes. Stations will incorporate amenities that are similar to light rail, with low-floor boarding platforms, next-bus arrival information, and prepay fare machines. The Mason Corridor has been recognized as an innovative "Small Starts" project by the Federal Transit Administration. Project construction is scheduled to begin in 2010 and be completed by the end of 2011. The city is seeking 80 percent of the total project costs from the federal government, with the other 20 percent from Colorado's Strategic Transit Program Funding (detailed in Senate Bill 1), the city, and the Downtown Development Authority.

One of the major successes of the Mason Corridor project is that it developed out of community discussions on transportation. The idea was studied by the city and the community and has grown into something that could serve as an example for other communities nationwide. Other successes include the transit-oriented development (TOD) overlay district that has been defined along the length of the corridor (developers are planning projects in advance of the BRT), the provision of needed pedestrian and bicycle grade-separated crossings over the railroad tracks, and a strengthened relationship with the BNSF Railway and CSU.

Research shows that Americans living in compact, well-planned communities where cars are not the only transportation option drive one-third fewer miles than those living in typical "sprawl" development. With its extensive and innovative transportation options, Fort Collins is a great example of how to achieve this type of VMT reduction and thus substantial GHG emissions reductions as well. ◄

ingly, the answer to the capacity question depends largely on when charging occurs and whether more of it can be done at off-peak hours. A national study by Pacific Northwest National Laboratory suggested that if electric vehicles commanded 15 percent of the market, existing generation and transmission capacity could absorb the demand, if timed right. If all the demand came at peak hours, however, 160 new power plants would be needed (Scott et al. 2007).

Parking

Providing priority parking spaces for hybrid and alternative fuel vehicles may serve to encourage the purchase and use of such vehicles. Likewise, creating standards for small parking spaces for neighborhood electric vehicles (NEVs) and other small cars shows support for this pollution-reduction and climate change mitigation technology.

Figure 8.8. A hybrid car powering up

Fleet Purchasing and Management

Government and corporate fleet purchasing programs can be important catalysts to stimulate demand for advanced vehicles. How communities manage their fleet of public vehicles can have a significant effect on greenhouse gas emissions. Concord, North Carolina, for example, has taken steps to manage its vehicle fleet in an energy-efficient manner. Concord's garage superintendent David Nuckols offers the following practical ideas for fuel-efficient fleet management:

- Purchase hybrid cars and small pickup trucks for most normal-use light vehicle situations.

- Equip full-size pickups with the smallest V-8 engine offered in the manufacturer's line.

- Purchase diesel engines for vehicles over three-quarters of a ton.

- Specify the most fuel-efficient diesel engine for large trucks; include in the specifications low horsepower ratings coupled with transmissions that provide adequate power at low speeds and, as necessary, the ability to reach highway speeds.

- Adopt idling policies that mandate turning vehicles off, while providing exceptions such as idling in traffic, to keep interiors cool for K-9 units, in extreme cold weather operations, etc. (Lail 2007).

TOOLS FOR GUIDING TRANSPORTATION DECISIONS

Visioning and Goal Setting	Assess the community's desire to be able to travel more frequently by alternative transportation modes and include goals for improving the transportation network to more safely and conveniently accommodate alternative modes
Plan Making	Update comprehensive plans, neighborhood plans, and transportation plans to allow for complete streets and infrastructure that supports multiple modes of travel
Standards, Policies, and Incentives	Revise street design standards to accommodate bicycle lanes, street connectivity, and appealing sidewalks Revise development standards and regulations to provide incentives for TODs and transit-supportive density, as well as to provide preferential parking for bicycles, NEVs, and other alternative-fuel and transportation-pool vehicles
Development Work	Review development plans for street connectivity, access, and similar considerations
Public Investment	Make public investments in infrastructure that allows travel by energy- and GHG-efficient modes

CHAPTER 9

Economic Development

 The impacts of energy and climate change issues on economic development may prove to be among the most significant issues of concern in the years ahead for local planners. The U.S. economy is extremely diverse, which may help buffer it against projected economic impacts from climate change. However, the U.S. economy is also interdependent with the global economy, and this interdependence will likely increase vulnerability to changing markets, resource limitations, and other global factors influenced by climate change. However, efforts to promote energy efficiency and renewable energy sources and to adapt to climate change impacts also present important opportunities for local economic development. This chapter reviews the economic development challenges and opportunities that planners may face in the following areas:

- Economic Migration
- Agriculture, Forestry, and Fisheries
- Shipping and Freight
- Manufacturing
- Insurance
- Services and Construction
- Tourism and Recreation

Additionally, strategies for adapting to economic development impacts are presented.

ECONOMIC MIGRATION

International trade is not typically a focus area for local planners. It is important, however, for planners to have an understanding of its impacts on their local economies in order to identify vulnerabilities that may be exposed as a result of climate change. In many cases, local economies have global connections due to significant changes in the structure of the U.S. economy over the last several decades. Expanding international markets, enhanced shipping technologies (e.g., containerized shipping and the development of "just-in-time" delivery), and relatively low energy costs have all contributed to the movement overseas of certain industries. As a result, manufacturing and other economic sectors have become less constrained by factors such as geographic advantage, transportation costs, or the availability of particular resources or labor markets. Accordingly, local economies are much more subject to global forces, as it becomes easier for businesses to move elsewhere when it is makes sense for them to do so (McLean and Voytek 1992, 133–34). As has been widely noted (e.g., Florida 2002), the "new economy" is highly mobile. Energy and climate change issues are likely to become increasingly important forces that may drive industries and people to migrate. Water supply problems, difficulties in finding affordable insurance, availability of cheaper energy and shipping, and other factors may cause businesses and industries to relocate to more advantageous places. Migration, therefore, has enormous implications for local economies, creating different issues depending upon whether business is moving in or moving out. Planners will need to develop adaptation responses to assist their communities as these economic transitions occur.

AGRICULTURE, FORESTRY, AND FISHERIES

Higher temperatures may mean longer growing seasons in some areas, increasing yields. Temperature changes may also increase the commercial range of certain crops, such as the ability to produce oranges in northern Florida. Higher CO_2 concentrations may spur enhanced plant growth, also increasing yields. These geographically based gains may be offset by problems from increased heat-induced evaporation, greater incidence of drought, more intense flooding, and increased pest activity. Issues for planners include addressing the following:

- Adjustments to crop type selection and management in response to new climate circumstances. These may result in new or different markets and the need for different or retrofitted machinery for planting and harvesting. Farmers may require assistance in making these adjustments.

- Balancing agricultural irrigation needs with the needs of other water users.

- New or more frequent pest management applications, including managing their effect on surrounding nonagricultural properties. As residential development into exurban areas occurs, there is potential for increased conflicts with or concerns about agricultural practices.

Drier conditions are expected to create more wildfires, reducing timber yields. Pests' ranges may increase along with their period of activity, also reducing yields; this is already occurring in many areas affected by the pine bark beetle, for instance. The need for enhanced fire management may limit forest production during and for some time after controlled burns. There may be declines or increases in the commercial yield potential for certain tree species as their historic ranges are altered by higher temperatures. On the other hand, greater CO_2 concentrations may spur increased growth, which could enhance yields.

Higher temperatures and changes in the salinity and pH of ocean waters may negatively affect yields from certain species of fish and shellfish. Sea-level rise, drought, and flooding may affect the spawning runs of salmon. Loss of wetlands due to sea-level rise, flooding, sedimentation, and erosion has the potential to affect many species negatively. Some fishing ports and processing facilities may be threatened by sea-level rise. Various studies note that thermal expansion of the oceans has the potential to affect major ocean currents, creating significant changes in fish populations and their migratory habits.

SHIPPING AND FREIGHT

Significant changes in shipping routes may occur as the Northwest Passage and the Northern Sea Route are able to be used more frequently as a result of declining sea ice. These new routes cut 5,000 or more miles off the distance traveled between Europe and China relative to the Panama and Suez canal routes. They have the potential to greatly lower shipping costs and are able to support the much larger container ships currently being planned or constructed. These routes may generate new activity in northern ports while reducing activity in southern ports, a potentially significant change for those local economies.

Sea-level rise will increase the vulnerability of many ports to inundation and storm surge. Similarly, road and rail access to ports and to coastal areas in general may change on account of problems associated with sea-level rise. An analysis of these impacts in the Gulf Coast region noted that a relative sea-level rise of four feet would permanently flood 72 percent of freight and 73 percent of nonfreight facilities of the region's ports (CCSP 2008).

MANUFACTURING

Water-supply issues may affect manufacturing operations and costs. Other problems for this economic sector include flooding or the inundation of manufacturing facilities located in vulnerable locations, weather-related raw material and finished product delivery issues, and the effects of higher temperatures on products and personnel, particularly for outdoor storage and processing operations.

INSURANCE

The insurance industry is expected to be significantly affected by climate change. The Global Business Network (GBN) notes that "insurance is enormously important, in part because without the socialization of risk, development becomes much more difficult" (Gilman, Randall and Schwartz 2007, 13). This source also describes three major challenges for the industry:

- "Insurance prices for events like floods, droughts, wind, hurricanes, and tornadoes are all based on historical data. . . . Climate change makes this historical data much less useful. . . ." As a result, some markets may prove overly risky, resulting in insurers exiting the market. Government may be forced to step in to fill the gap in order to maintain both public and private investment, such as in the case of federal flood insurance. This step would, of course, expose the public sector to risks that the private sector is unwilling to endure.
- Regulators may hold the actual prices of insurance to artificially low levels due to concerns about the effects of actuarially based pricing using climate projections on, for example, the real estate market in hurricane-prone areas. As with the uncertainty issue, this has the potential to affect the willingness of insurers to remain in higher-risk markets and to result in government action if private insurers demur.
- The potential for increased exposure to climate change liability lawsuits further complicate risks. For example, GBN postulates a class-action law-

suit against a corporation that produces 1 percent of human-induced GHG in which the corporation is held accountable for 1 percent of the global damages associated with climate change. (Gilman, Randall and Schwartz 2007, 13-14)

SERVICES AND CONSTRUCTION

Health care will need to adapt to the increasing trend in heat-related maladies, vector-borne diseases, and the need for disaster response services. Outdoor-oriented businesses will have to adjust to warmer temperatures, which may create heat-related delays in some areas and a prolonged outdoor work season in others. New markets for energy-efficient, green building products and construction techniques will open to address climate change issues such as water-supply problems caused by drought or contamination and cooling costs caused by higher temperatures.

TOURISM AND RECREATION

Rising temperatures will affect cold-weather tourism and recreation by reducing amounts of snowfall and the length of the winter-weather season. In many areas, lake levels will be reduced by drought and evaporation, complicating such recreational activities as water sports and fishing. The GBN has identified the following problems for beach resorts:

> Warmer winters in North America and Europe will decrease demand for sunny winter escapes. At the same time, the "beach product offering" is likely to become less attractive as the heat index rises, beaches erode, sea and coral quality decline, and vector-borne diseases increase. (Gilman, Randall, and Schwartz 2007, 14)

In December 2007, the *New York Times* reported on efforts of European ski resorts to diversify their offerings in response to climate change:

> With glaciers melting and snowpacks shrinking, ski resorts in the Alps are trying to stay ahead of global warming, not only by installing more snow-making guns, but also by transforming their resorts with colossal spas, sleek architecture and other off-slope attractions. Big-name architects like Zaha Hadid are designing high-altitude ski features. Shopping centers are going up on mountain peaks. And venerable hotels like the Tschuggen Grand are becoming all-weather resorts, in its case by adding a $30 million, 43,000-square-foot spa designed by the Swiss architect Mario Botta. (Williams 2007)

ADAPTING TO ECONOMIC DEVELOPMENT IMPACTS

Climate change mitigation responses involving physical planning that affects local economic development center largely on three approaches, all of which have a transportation focus. The first approach attempts to integrate business and industrial activity into the urban fabric in order to promote more compact development. The second approach ensures that there are multimodal transportation options to reach places of work. The third approach uses transportation demand techniques to either reduce the amount of work-related travel or shift it to off-peak travel times to reduce congestion that makes travel less efficient (see Chapter 8).

Since commuting to work is a major component of overall U.S. vehicle-miles traveled (VMT), reducing driving distances through compact development patterns, encouraging transit, bicycling, and walking modes of commuting, and using transportation demand management to reduce trips or make them more efficient are obvious opportunities to mitigate GHG emissions.

Figure 9.1 illustrates VMT in a major metropolitan area (Detroit) over a 28-day period. The diamonds on the chart indicate days of the week. Sundays see the least travel, while weekdays are relatively consistent, with modest increases toward the end of the week as more discretionary travel occurs.

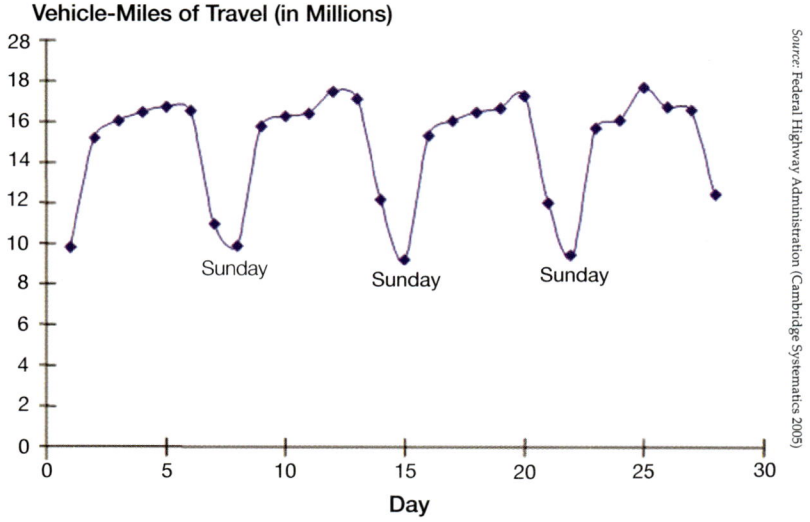

Figure 9.1. Traffic levels on Detroit freeways, March 11–April 7, 2001

Commuting patterns are generally consistent during the weekdays and clearly represent a major component of VMT.

Compact Development Approach

Compact development patterns offer the opportunity for employees to use alternative transportation methods to get to work. There are a variety of approaches that can be used to integrate employment into a compact urban form. Mixed use development, especially when it includes nonresidential uses beyond the typical retail, restaurant, and neighborhood services uses, can increase the chances for nearby employment. Home occupations and live-work units can effectively integrate residential and employment uses. Employer-provided housing can do likewise. Buffering and separation requirements in local codes can be examined for continued validity given improvements in industrial processes involving noise, site activity, and emissions. New zoning districts can be created to allow light industrial uses in greater proximity to residential development. An "employment density map" can be developed to identify areas of highest employment in a community, allowing planners to target infill development, density incentives, transit services, and other land use and transportation measures to more effectively link employment with residential development. New use categories can be created to help incorporate certain traditionally industrial uses into a broader range of zoning districts, as in this example:

> *Research and technology production uses* means uses such as medical, optical and scientific research facilities, software production and development, clinics and laboratories, pharmaceutical compounding and photographic processing facilities, and facilities for the assembly of electronic components, optical equipment, and precision instruments. (Asheville 2009)

In general, planners should examine their code requirements to determine if they are overly restrictive concerning the integration of industrial and other employment uses into the overall community land use pattern, initiate conversations with major employers to determine their willingness to accommodate or support on-site or nearby employee housing, and map employment density in order to create a comprehensive program of "densification" around centers of employment. Additionally, industrial recruitment efforts can be tailored to attract industries which are compatible with local land-use patterns.

Multimodal Transportation Approach

The employment density map discussed above is a good place for planners to start in developing a program to encourage multimodal transportation options to and from employment centers. Provisions can be included in local regulations to encourage or require transit, bicycle, and pedestrian access to major employment centers. Transit stops and transit-oriented developments (TODs) can be sited to serve employment centers, and areas where these transportation facilities already exist can be zoned to encourage the nearby location of industries and other employment uses. Employers can be approached about participating in the provision of multimodal transportation facilities to give their employees greater choices about how they get to work; in some cases, development approvals can be conditioned on the provision of such facilities. Existing or planned multimodal transportation facilities can be used effectively in marketing the economic development potential of communities to certain industries.

Planners need to think comprehensively about how their communities' transportation systems can support economic development while reducing VMT.

CLIMATE CHANGE ADAPTATION AND ECONOMIC DEVELOPMENT

Adaptation responses to climate change involving local economic development should address vulnerabilities in two main areas and adjust to in-migration of industry and business in areas where such vulnerabilities are minimized. The two main areas are:

- *Physical vulnerabilities:* Industrial facilities, agricultural operations, ports and shipping facilities, power generators, and other support infrastructure may be located in areas prone to inundation due to sea-level rise, in areas made more floodprone due to changes in precipitation patterns and snowpack melting rates, and in areas of thawing permafrost. Consequently, there are physical threats to the local economy that must be addressed by relocation, renovation, or innovation to eliminate the threats or minimize them to acceptable levels.

- *Sector vulnerabilities:* Some sectors of the economy may be threatened by changes in resources available as products or for production or processes. For example, drought-related water-supply issues may affect availability of water for use in industrial processes, for hydroelectric generation, or for irrigation. Certain types of agriculture may not be feasible in drier conditions. Drought may also affect tourism through reduced lake levels or wildfire-charred scenery. In these sector cases, planners may need to address vulnerabilities through innovations in processes, practices or energy use, land-use conversions or adaptive reuse of existing facilities, specialized or time-sensitive marketing, and similar measures.

Addressing physical and sector vulnerabilities will require planners to clearly identify the specific risks faced by the local economy. Do stronger electrical storms pose particular threats to the broadband cable network that supports the new high-tech company in town? Will the proposed timing of a controlled burn affect the scheduling of a particular tourism event? Will lower lake levels cause the marina to close? If sea-level rise inundates the port, what additional businesses will be affected?

This risk assessment should be supplemented by cost-benefit study or an opportunities analysis that evaluates the effects of possible adaptation measures. Tourism, commercial fishing, and recreational fishing may be enhanced by the removal of a hydroelectric dam made obsolete by climate change due to low water flows, for instance. Such studies also need to take into consideration uncertainties in climate change impacts at regional levels so that responses are balanced with risk potential.

Physical Vulnerability Adaptation

Coastal and floodprone areas and areas facing permafrost thaw are some of the most physically vulnerable locations for climate change effects. However, any area in the United States may experience stronger tropical and nontropical storms, drought conditions, higher temperatures, wildfires, or other climate change impacts that may create physical vulnerabilities for different segments of the local economy.

Physical vulnerabilities will create a range of economic development responses and impacts:

> **Migration**: the use or economic activity can no longer effectively function in its current location and migrates from the locality or region in response to climate change.
>
> **Obsolescence/abandonment**: the use or economic activity can no longer effectively function in its current location and becomes obsolete or is abandoned in response to climate change.
>
> **Relocation**: the use or economic activity can no longer effectively function in its current location but can be relocated elsewhere in the community or region in response climate change.
>
> **On-site mitigation**: the use or economic activity can continue to function in its current location provided impact mitigation measures are taken in response to climate change.

In the event of migration or obsolescence/abandonment, planners will need to take steps to address the effects of the loss of the particular economic use or activity. With regard to relocation and on-site mitigation, planners will need to determine what level of governmental participation can and should be provided in accommodating the resultant costs. This determination will depend on the importance of the economic use or activity to the local economy, the scale of private-sector impact, the government's ability to pay, and other factors.

Sector Vulnerability Adaptation

Sector vulnerability responses to climate change impacts are similar to those for physical vulnerabilities. However, there are typically more options for communities to pursue, since buildings housing obsolete uses can be adaptively reused or new crops can replace those that are no longer viable. The response categories are listed below, with additional discussion relevant to sector vulnerability issues and opportunities.

> **Out-migration:** the use or economic activity can no longer effectively function in its current location and migrates from the locality or region in response to climate change. This may be the case of a water-dependent use in a drought-prone location. In most cases, however, the facilities and infrastructure remain intact and available for reuse for another economic activity or use. The problem then becomes one of marketing the location to suitable alternative uses, rather than adjusting to a permanent loss of the resource as in physical vulnerability.
>
> **Obsolescence/abandonment**: the use or economic activity can no longer effectively function in its current location and becomes obsolete or is abandoned in response to climate change. In this case, the opportunity for adaptive reuse remains. There are numerous examples of abandoned wharves and similar facilities being converted to housing or mixed use developments. A word of caution is warranted, however: planners will need to evaluate the need to retain existing warehouse

and industrial sites in the event of the reemergence of a more decentralized model of freight distribution and local production due to rising energy costs and other factors.

Relocation: the use or economic activity can no longer effectively function in its current location but can be relocated elsewhere in the community or region in response climate change. Relocation offers the opportunity to adaptively reuse the facility or property.

On-site mitigation: the use or economic activity can continue to function in its current location provided impact mitigation measures are taken in response to climate change. Such measures may include switching to alternative crops or agricultural practices, floodproofing, hardening the water's edge, and modifying the support infrastructure to create climate change resilience.

Adaptation to in-migration: Because of favorable location relative to climate change impacts or because of proactive efforts to create climate change resilience, some communities will find themselves in the position of having increased economic development pressure because businesses and industries will be migrating to them.

While this may appear to be a good problem to have, rapid development can strain infrastructure capacities, create congestion, and affect quality of life; additionally, if not properly accommodated, such growth can exacerbate climate change impacts by increasing the location's carbon footprint. Therefore, it is important for planners to attempt to accommodate this growth in a compact urban form, with multimodal transportation options and an appropriate range of transportation demand management tools.

In some cases, the economic development opportunities may be related to agriculture or forestry as the commercial growing range expands for certain crop species or if there is available water for irrigation. As with urban businesses and industries, planners need to ascertain the impacts associated with changes in agricultural practices from new crops; for example, exchanging a forestry-based economy for a farming-based economy may have associated economic development impacts that may affect the desirability of the exchange. Other industries, such as tourism, may be better supported by a reliance on the forestry sector than by a reliance on the agricultural sector—tourists may prefer to look at trees rather than fields, for example. Such factors need to be taken into account in promoting changes to existing economic development practices.

Planners in areas likely to experience enhanced economic development as a result of climate change would be well-served in developing a sustainable economic development plan that identifies focus industries and green economy opportunities. Having such a plan will enable those communities to more effectively market themselves because relocating industries and businesses will understand and appreciate the need to introduce sustainability and climate resilience into the local economic development picture if for no other reason than the climate change impacts that prompted their relocation.

ECONOMIC DEVELOPMENT OPPORTUNITIES

At the same time that climate change and energy issues present numerous economic challenges, they also create new economic development opportunities. Economic development opportunities resulting from climate change responses, some of which have been noted by the United Nations Environment Programme (UNEP 2008), include:

- Growing numbers of companies are embracing environmental policies, and investors are pumping hundreds of billions of dollars into cleaner and renewable energies

- The emerging green economy is driving invention, innovation, and the imagination of engineers

- Combating climate change is increasingly being perceived as an opportunity rather than a burden and a path to a new kind of prosperity

- Many companies now perceive that going green improves their bottom line

- Companies in six sectors—ranging from mining and energy to food and media—that have adopted pioneering environmental, social, and governance strategies are outperforming the general stock market by 25 percent

These trends have clear implications for local economic development. Due to consumer preference and regulatory requirements, there is a growing demand for green products and services, such as locally produced organic foods, environmentally-sensitive construction techniques, and energy-efficient vehicles and appliances. There are huge opportunities to mitigate climate change through building and site design and to develop a local economy around green building products and techniques. Rising energy costs will create markets for energy-efficient products. New vehicle technology can use locally produced biofuels or locally generated or co-generated electricity. Training workers in green technology can transform local labor forces.

Economic development should be geared to enhance the resiliency of local communities to weather the economic changes and uncertainty that climate change and energy transitions may bring. Economic development also presents a critical opportunity to reduce energy use and greenhouse gas emissions while strengthening the local economy.

The Climate Prosperity Project—an initiative involving the International Economic Development Council (IEDC) to help economic development professionals in the United States and abroad develop regional "climate prosperity" strategies—identifies three key areas of economic development opportunity:

- Energy Cost Savings for Businesses
- New Market Opportunities
- Workforce Development

IEDC uses the terms "Green Savings," "Green Opportunity," and "Green Talent" as shorthand for these concepts. An important theme of this work is that there is not a special part of the economy that is the "green economy" nor a special class of "green jobs." The opportunities are not confined to one or a few industries related to energy production but rather extend to all energy users, which means the entire economy. Every industry, business, and job can save energy, shift to renewable energy, and reduce GHG emissions. Similarly, every company would be wise to consider how climate change may affect their business.

ENERGY COST SAVINGS FOR BUSINESSES

Many energy efficiency and renewable energy measures will pay for themselves and save money in the long run—and often in the short run, too. Energy and climate issues are providing the impetus for many businesses

MAKING IT GREEN IN MINNEAPOLIS–ST. PAUL
Population: 646,368

Making It Green is a collaborative effort to recommend policies and actions that will position Minneapolis and St. Paul as national leaders in promoting and developing green manufacturing technologies and products, while at the same time creating family-supporting jobs. The report is an initiative of the two mayors of the Twin Cities, along with the Blue Green Alliance, a national partnership of the Sierra Club and the United Steel Workers. It highlights the growing recognition that cities and regions that invest in technology and workforce education programs will have an edge in solving critical environmental challenges, such as global warming. Building a "green economy" gives cities strategic economic opportunities for the future.

The study recommends five strategies: aggressive marketing, realigning economic development tools to focus on green industries, growing markets for local suppliers, strongly lobbying for state policies and incentives to support green job creation, and forging enduring partnerships between the cities by continuing ongoing work on the Mayors' Green Manufacturing Initiative. The study found that revitalizing Minneapolis and St. Paul's economic base will help capture some of the national green manufacturing economy that is currently valued at $229 billion.

Figure 9.2. A photovoltaic module manufacturing facility

The Twin Cities is the fourth-largest manufacturing area in the United States, with both cities supporting policies that attract industrial firms. The region possesses ready marketplaces for green products and suppliers and enjoys an increasing variety of renewable energy sources and lower-impact transportation and shipping options, such as via the Mississippi River.

In researching and writing the Making It Green report, the authors looked at three green manufacturing sectors with high potential for near-term growth and job creation: green buildings, renewable energy, and transportation industries. They identified a substantial base of manufacturers with product applications for 29 subsectors of these industries. The report focuses primarily on manufacturers in these sectors that produce goods that reduce GHG emissions and that can provide immediate employment opportunities. In future phases of the project, there will be efforts to encourage other green industry sectors, as well as efforts to increase the "greening" of other manufacturers.

The report also provides a survey of green economic development best practices in cities across the United States. The survey includes examples of the three-step strategy the Mayors' Initiative recommends:

▶ Leading by example with commitment and focus from the top

▶ Making markets by supporting demand for green products

▶ Adopting green business as an economic development strategy

The research presented in the report represents phase 1 of the Making It Green campaign; the next phases will focus on developing strategies to achieve the campaign goals and then implementing them. John Dybvig, director of economic development for the Blue Green Alliance, reports that the alliance, along with representatives from the Twin Cities, went to the State of Minnesota and secured funding to act on the recommendations from the report. Minneapolis and St. Paul are also both contributing stimulus funds to a joint effort for energy-efficiency retrofits.

Some of the current implementation activities are centered on identifying green products already being manufactured locally and the markets for those products. Dybvig further states that the organization is studying "what green manufacturers would like in the form of regulatory environment and where they see the gaps in the current system. We are also developing a marketing plan to make sure it is well known that we are open for green business."

The report is already having an impact. Minneapolis is currently working with a local solar PV manufacturer that is scaling up its operations. The city is proposing to use a portion of its Community Development Block Grant Recovery (CDBG-R) funds, allocated as part of the 2009 federal stimulus package, to assist this company with build-out costs for a new facility. Since solar PV manufacturing had been identified in the report as a desirable industry, it was easier to garner political support for this allocation.

Switching local, regional, national, and transnational economies over to more environmentally sound practices will be critical to limiting GHG emissions and both mitigating and adapting to climate change. Planners will be able to facilitate these efforts by taking the lead in economic development initiatives like the Making It Green campaign.

to capture these savings. Businesses can create considerable cost savings by identifying hidden efficiencies available to existing operations, seeking assistance from outside experts, and seeking innovative ways to finance long-term investments in energy efficiency.

Hidden Efficiencies

Opportunities for greater efficiency can be hard to see because many of them are small and diffuse, but once realized they can add up to substantial cost savings. For businesses, efficiencies can be spread across general building operations, transportation, or activities and processes specific to an individual business. For example, the 3M Company employed more than 1,000 separate measures on its way to improving energy efficiency by 77 percent since 1975 and reducing GHG emissions by more than 60 percent since 2002 (U.S. EPA, Climate Leaders. n.d.).

More than half of these measures paid for themselves within five years, most of them did so within 10 years, and they continued to yield savings thereafter. As the company planned to be in business for the long term, the benefits were clear.

Efficiency measures are especially hard to see if people are not looking for them. Many companies have reported success just by establishing energy efficiency as a performance goal. An important element of success for many businesses has been the engagement of employees who are most familiar with the details of daily business operations. Companies such as Coca-Cola, IBM, and Johnson & Johnson—who participate in Climate Savers, a program for large corporations spearheaded by the World Wildlife Fund—credit their employees with the success of their efforts. Some companies in this program also note the importance of incentives, such as allowing department managers to maintain discretion over how the cost savings they achieve will be reinvested (EESI and WWF 2009).

Outreach and Technical Assistance

While involving internal staff is critical, many efficiency measures may be more readily identified by persons with specialized training, experience, and equipment to assist with comprehensive energy and GHG emission audits and ongoing improvement programs. Many businesses report that the effectiveness of their efforts was catalyzed by a partnership with an outside entity, such as the EPA's Climate Leaders program and the WWF Climate Savers program (EESI and WWF 2009). Other nonprofit organizations such as the Environmental Defense Fund also have strong corporate partnership programs. Smaller companies may choose to work with state or local programs. For example, the Vermont Energy Investment Corporation provides energy services to businesses statewide. In other cases, an outside consultant or a targeted program of a local utility or government agency can be instrumental in creating an enduring program.

While the financial benefits are clear and demonstrable, the participation in such a partnership can itself be an intangible asset to the culture and public image of the company, demonstrating community involvement and corporate responsibility.

Financing

No matter how clear the cost benefits are, companies need a way to pay for energy saving and GHG emission reduction measures. Companies may not have immediate surplus capital to invest in what they see as "extras." Many businesses are also uncertain about their own longevity or locations. The availability of financing instruments designed especially for businesses can greatly accelerate the deployment of energy efficiency and renewable energy measures.

Financing assistance is best coordinated with an internal or external technical analysis that compares the investment return or payback period with the cost implications of different energy options. Many cities and states have set up a revolving loan fund that allows businesses to pay down loans for energy measures with the cost savings that such measures achieve. These programs help resolve uncertainties and barriers regarding financial liability and ownership of the assets should the business move or close. An energy services company (ESCO) may be able to develop, install, and finance an energy efficiency project, whereby the energy savings pay for the project.

NEW MARKET OPPORTUNITIES

Efforts to reduce energy use and GHG emissions across all sectors of the economy do not help only the financial bottom line of an individual business; they also provide a large and growing market development opportunity for all businesses. As noted, energy and GHG savings can be found among virtually all businesses. This creates countless opportunities for other businesses connected to them, upstream and downstream, to provide more low-energy and low-carbon products and services. Moreover, these are not one-time opportunities, as competitive forces are expected to drive perpetual improvement in the efficiency and cost-effectiveness of energy and cost-saving measures. The green market trend started narrow and shallow, but it is becoming deep and wide.

Growth Trends

An economic development strategy focused on energy, climate, and other "green" parameters is not just a side benefit of altruistic action on energy and climate issues. Trends suggest that it may be the dominant long-term growth trend of the 21st century. Growth in the number of narrowly defined "green jobs" and businesses has outpaced growth in just about every sector of the economy other than health care.

Over the long term, the U.S. economy has been shifting away from manufacturing and exports and toward services, imports, finance, and consumer products. However, the economic downturn that began in 2007, led by a collapse of the housing bubble, may be shifting the economy away from a consumption-led economy and back toward savings and investment. Meanwhile, less developed but industrializing countries are investing strongly in infrastructure and primary industries as they strive toward economic development levels comparable to those in industrialized countries. Overall, products and services that are integral to economic prosperity are likely to have more sustained growth potential than peripheral and nonessential products. And nothing is more integral to economic prosperity than energy.

Stimulating Demand

Any new business needs to build a customer base for its product or service. Once this foothold is established, demand for one line of products may stimulate demand for related products, building broader industries. Growing demand for insulation, for example, spills over to related products such as energy-efficient windows, programmable thermostats, and so on.

More important, once a significant level of demand is demonstrated for one class of products, other businesses start to compete for that business, propelling innovation and competition in that market. Catalyzing that initial level of demand can be a key to unleashing the market development process. Hence, one major purchaser of goods and services—a large corporation, a university, or government agency, for instance—can play a critical role in realizing latent market opportunities.

Demand for energy products and services can build industries that use domestic production to satisfy domestic consumption. Despite the unavoidable dynamics of a global marketplace, many businesses find it prudent to focus first on satisfying domestic demand as a countermeasure against possible fluctuations and drop-offs in overseas demand. It still can be important and profitable to sell products and services for an export market, but the U.S. market can easily create initial market demand. Also, energy and GHG emission reduction–related business opportunities tend to be products and services that are not easily outsourced.

Economic Diversification and Localization

As new market opportunities related to energy and GHG reduction develop, economic diversification and localization are trends to which planners may need to give renewed attention. New business and employment opportunities related to energy and GHG reduction may not immediately replace the tens of thousands of jobs that have been lost in some industries. It is unlikely that any one industry could do so. But the new energy economy is not just one industry—it is all industries cast in a new light and working within a new economic model of efficiency and renewable energy.

As with prior economic development models, diversification is still an important objective. Reliance on any one industry, no matter how green or how perpetual the demand may seem to be, is not wise. Diversity will be particularly important as an adaptation strategy as well. As changing energy markets and climate change exert new forces on business relations, understanding which industries are more or less vulnerable to supply-chain disruptions or price shocks due to climate change is an important part of assessing a community's economic resiliency.

Individual businesses as well as regional economies are likely to be forced to diversify the markets they serve. Many ski areas, for example, used to be one-season operations, but warmer temperatures and less consistent snowfalls have led to expansion into other seasons.

In an increasingly global economy, it may seem outdated to suggest that communities look to businesses closer to home to meet their needs, but changing energy and climate factors strongly suggest that may be necessary. Potential price increases for energy, both directly and for other businesses in a supply chain, may substantially alter the profit margins for goods that involve transport over long distances.

Moreover, the production of raw materials and other intermediate inputs may be disrupted by one or several climate impacts, threatening the long-term reliability of existing supply chains. It is a good idea to develop supply chains that diversify at least some of their sourcing to include local or regional sources where possible. This is especially true for agricultural products, which may be susceptible to extreme weather events, high transportation costs, and long-term climate trends.

This potential for significant impact on local economies can be realized by adopting a comprehensive approach that promotes synergistic relationships among local industries, institutions, and businesses. For example, Asheville, North Carolina, is developing a "Sustainability Institute" that will "research issues like development, transportation, recycling and alternative energy in an attempt to turn the concept of sustainability into new jobs" (Neal 2008). This effort would build on the work of the Asheville Hub Alliance, a nonprofit economic development organization that was created to facilitate interaction among the region's economic sectors of health care, the arts, technology, education, climate research and services, entrepreneurship, and tourism.

PIONEER VALLEY ENERGY PLAN
Population: 687,658

Figure 9.3. The Connecticut River valley, in Massachusetts

The Pioneer Valley Planning Commission (PVPC) has developed an energy plan for Hampden, Hampshire, and Franklin counties in western Massachusetts that can serve as an example to other areas hoping to develop a path forward to a new energy future. Catherine Miller of the PVPC says, "We made a serious and meaningful and successful effort to include as many people, organizations and individuals as wanted to be included in the development of the plan, [and] the goals of the plan really are exactly what we all need to be doing—we highlight efficiency and conservation as the top priorities; we do not shy away from the need to generate new power; we aggressively address the need to reduce greenhouse gas emissions and we understand that in meeting the first three goals we will create good green jobs."

During the three-year planning process, the plan's authors found that there was overwhelming support throughout the region for developing clean energy sources and reducing emissions that cause global warming. The PVPC harnessed that support to develop an energy plan that focuses on four comprehensive short- and long-term goals: energy consumption reduction; clean energy generation; GHG emissions reduction; and local job creation in the clean energy sector.

The Pioneer Valley has been a manufacturing center since the Springfield Armory produced weapons for General George Washington. Since then, the region's hydropower and inexpensive labor has facilitated the production of paper, textiles, and machined goods. But in recent decades, the manufacturing base declined steeply as industries changed and restructured. The industrial corridor along the Connecticut River is dotted with shuttered mills and factories.

The PVPC would like to see a return to the manufacturing roots of the region, but with an eye toward energy-efficiency technologies and clean energy production. Research by the Massachusetts Technology Collaborative (MTC) indicates that the clean energy sector is about to overtake textiles as the 10th-largest industry group in the Commonwealth. Rather than depend on recruiting big companies from outside the region, the energy plan suggests that the region leverage its concentration of colleges and universities in order to develop locally based small businesses that benefit from research conducted at local schools.

Public outreach during the planning process found that residents of the valley felt strongly that most of the renewable and clean energy projects should be developed by local entities, whether through municipalities, local nonprofits, locally owned businesses, or cooperatives. Residents have seen companies come and then leave with devastating effects, and they do not want that to happen again. The plan points out research showing that local businesses represent a better option for building local economies not only because they are vigilant about maintaining high labor and environmental standards but because they keep profits in the community, spending two to four times more in the area than nonlocal companies do.

To help attain the goal of clean energy job creation, as well as the other energy plan goals, the plan contains a set of guiding principles and criteria for project selection and development. The principles and criteria were developed with the help of participatory planning processes and focus on reducing energy consumption and fossil fuel use while increasing small-scale renewable energy projects that create well-paying local jobs, provide affordable energy for everyone in the region, and make the Pioneer Valley a leader in clean energy technology. The criteria also encourage more compact land use, better public transportation options, and the retention of privately owned open space by siting carefully chosen renewable energy projects on suitable parcels.

Figure 9.4. The PVPC plan calls for focusing on small-scale renewable energy projects like the biomass gasifier located at the McNeil Station in Burlington, Vermont, which can convert up to 200 tons of biomass per day into a clean-burning gas fuel for power generation.

The PVPC encourages its constituent communities to work both individually and collectively to increase policy and regulatory efforts that will make siting, financing, permitting, code compliance, and other project tasks easier and quicker. Many

(continued on page 115)

While Asheville has some unique characteristics that support such an approach—notably institutions such as the University of North Carolina at Asheville and NOAA's National Climatic Data Center, as well as major health care and tourism sectors—virtually any community has the opportunity to capitalize on the growing focus on "green collar" jobs that contribute to preserving or enhancing environmental quality. Such jobs range from the manufacture of green products like biofuels and solar panels to green construction to organic food production.

For example, a community may choose to create what Ernest Lowe of Indigo Development terms an "eco-industrial park" or EIP to take advantage of synergies between green businesses.

> An eco-industrial park or estate is a community of manufacturing and service businesses located together on a common property. Member businesses seek enhanced environmental, economic, and social performance through collaboration in managing environmental and resource issues.
>
> By working together, the community of businesses seeks a collective benefit that is greater than the sum of individual benefits each company would realize by only optimizing its individual performance.
>
> The goal of an EIP is to improve the economic performance of the participating companies while minimizing their environmental impacts. Components of this approach include green design of park infrastructure and plants (new or retrofitted); cleaner production, pollution prevention; energy efficiency; and inter-company partnering. An EIP also seeks benefits for neighboring communities to assure that the net impact of its development is positive. (Lowe 2001)

What Lowe proposes is a systems approach to siting industrial development: placing industries that use the by-products of other industries or that can share energy systems and other resources in close proximity; anticipating green construction and infrastructure in industrial park layout and design; and collaborating with the surrounding community for services or resources or to ensure compatibility, among other synergistic and environmentally friendly practices. The goal is to create a node of industrial sustainability that minimizes waste, enhances interindustry cooperation, and more effectively and efficiently utilizes local resources.

In addition to supporting these types of community-based or regional initiatives to create a sustainable green economy, planners can take other complementary steps in their local regulatory and economic development incentives processes, including:

- Creating development incentives (faster permit processing, reduced permit fees, economic development grants, etc.) for green industries and energy-efficient construction projects.

- Eliminating or revising zoning restrictions on local agriculture operations that utilize sustainable practices in a fashion that is compatible with surrounding land uses. At present, many communities prohibit virtually all agricultural uses in urban areas.

- Developing recruitment programs for green industries, including supportive infrastructure investment, transportation and land use policies, and inventories of existing industries' resource needs and by-products to promote synergistic industrial location and interaction.

WORKFORCE DEVELOPMENT

Outsourcing is a critical issue in today's global economy, as economic shifts have displaced many workers in the United States. Although

---

(continued from page 114)

types of renewable energy technologies are inadequately addressed in the region's zoning ordinances, which could be a significant barrier to timely development.

Miller notes that since the plan's debut in January 2008, 32 of the region's cities and towns have officially endorsed the plan; the PVPC is in the process of reaching out to the remaining 11. As of July 2009, only one community had decided against endorsement. The group that facilitated development of the plan, the Pioneer Valley Renewable Energy Collaborative (PVREC), continues to meet, and PVREC subgroups are involved in grant proposals and other projects intended to help move the region toward a clean, safe, sustainable energy future. The Hampden County Regional Employment Board used information contained in the clean energy plan while coordinating countywide applications for Massachusetts "Pathways Out of Poverty" funding, as it wants to generate good green jobs that reduce energy use. Additionally, the number of municipal energy committees in the valley has increased substantially. Meanwhile, other regions in the state are looking to the plan as a model for their efforts. All these developments indicate that the plan is being taken seriously and point toward continued implementation.

Taking stock of how well a plan is implemented is critical, and the plan provides measurement tools for this purpose. The PVPC will use the data the MTC collects to track how many green jobs are created and green businesses developed. They also plan to study the economic and environmental impact of local clean energy projects, as well as of clean energy projects owned by nonlocal companies, to determine best practices and future use of resources.

The development costs for the plan were funded by the Massachusetts Technology Collaborative Renewable Energy Trust. The trust also provided additional funding for both a PVPC planner and one at Franklin Regional Council of Governments, a plan codeveloper, to continue to convene the PVREC as the members work to secure funding to implement the plan's goals.

it is helpful to have an economic development strategy that is outsource resistant, change in the global economy is constant. Therefore, a strategy should focus on developing the skills, knowledge, and capabilities of all workers more than on protecting jobs. This means growing and developing "green talent." A specific job may be gone tomorrow, but talent stays with the individual.

Cultivating green talent means using existing skills in the workforce in new ways and classifying jobs not by industry but by skills, which are transferable. For example, the automobile industry developed a population of skilled tool and die makers. The skills those workers have are useful in other manufacturing operations. The same is true for companies as a whole. For example, Cardinal Fasteners in Eaton, Michigan, makes large precision bolts, which for a long time they supplied primarily to the automobile industry. Cardinal Fasteners now sells more than half of its products to the wind turbine industry.

Tailoring an economic development strategy in a given region requires recognizing and building on the inherent strengths of the existing regional economy. For example, when rubber manufacturing in northeast Ohio steadily moved overseas, industry and government leaders formed the Ohio Polymer Strategy Council. This group built on Ohio's world-class expertise and capacity in polymer research, development, and manufacturing, which had been gained through the rubber industry, and applied it to a wide variety of other industries.

According to some studies, a majority of jobs in the green or "new energy" economy will be traditional skilled positions such as construction, administration, and finance, while others will be new specialized technical positions (American Solar Energy Society 2009). Many green jobs require well-trained, technically proficient workers, with math, science, and analytical skills. Workforce development programs—from technical programs in high schools and community colleges to advanced degrees and continuing education programs at universities—can emphasize these skills through curriculum development. Understanding how these skills apply to the new energy economy can help jobseekers adapt their particular skills to different types of emerging industries.

SUMMARY

Maintaining a strong local economy will prove to be a significant challenge for many planners, due to the potential for extensive vulnerabilities that can result from energy sector impacts and from climate change. Local economic development can have a "green" focus, with sector concentrations in efficiencies, markets, and workforce. Climate change mitigation can be addressed through the creation of a compact urban form, a multimodal transportation network, and a sound transportation demand management program, all of which need to have an economic development focus. Development of a strategic and comprehensive green economy program offers the opportunity of enhancing the local economy while helping address energy and climate change issues.

Adapting to climate change from an economic development standpoint may involve permanent loss of infrastructure and facilities due to inundation, flooding, or resource supply concerns, requiring communities to adapt to economic out-migration. In other cases, physical and sector vulnerabilities will require adaptive reuse of existing facilities, relocation of certain businesses and industries, or on-site mitigation of climate change vulnerabilities.

Some communities will experience a net enhancement of economic development opportunities as a result of favorable location, resource availability, or good planning in addressing energy transitions or climate change. These communities may find it particularly helpful to have a sustainable economic development plan to rely on for evaluating the appropriate mix of industry and to serve as a marketing tool.

TOOLS FOR GUIDING ECONOMIC DEVELOPMENT DECISIONS

Visioning and Goal Setting	Set goals related to a clean-energy economy and projected climate impacts
Plan Making	Update economic development strategies in local plans and consider how industries may move and change over time in relation to energy and climate considerations
Standards, Policies, and Incentives	Create incentives for local businesses to become more energy efficient Look for ways to promote synergies among green businesses
Development Work	Consider how development occurring through public-private partnerships may help green the local economy
Public Investment	Encourage public investments in facilities and infrastructure that will help local economic sectors save money while reducing energy use and GHG emissions, as well as adapt to climate change

CHAPTER 10

Buildings and Site Design

In the United States, buildings account for approximately 39 percent of total energy use, 72 percent of electricity consumption, and 38 percent of CO_2 emissions (U.S. Green Building Council 2009). Consequently, adopting building and site design practices that are environmentally responsible and resource-efficient can play a major role in reducing energy use and environmental impacts associated with buildings. On average, high-performance green buildings that incorporate sustainable design, materials, and construction practices typically achieve overall energy savings of approximately 26 percent and generate 33 percent fewer GHG emissions than do comparable conventional buildings (U.S. Green Building Council n.d.). Such efficiency gains are ultimately where local building practices need to aim, especially as technology and cost performance improve. In addition to being more energy efficient, buildings and sites can be designed to accommodate renewable energy technologies that may be installed upfront or in the future. Site and building design can also enhance access to public transit, bikeways, and sidewalks, which can help reduce energy use and GHG emissions associated with transportation as well.

Climate change adaptation also needs to be considered in site plans and building practices and locations. In this regard, drier, hotter climates present one type of adaptation challenge, while flood and storm-surge issues present another. This chapter outlines some important building and site design considerations for planners and communities, including:

- Maximizing Renovation and Reuse of Existing Buildings
- Increasing the Thermal Efficiency of Buildings
- Structure and Site Hazards Resulting from Climate Change
- Landscaping and Site Design Issues
- Options for Lighting and Fixtures
- The Creation of Renewable-Ready Sites and Buildings

MAXIMIZING RENOVATION AND REUSE OF EXISTING BUILDINGS

While older buildings might not be as energy efficient as buildings constructed today, they contain a great deal of embodied energy—that is, the energy it took to make their materials, transport them to their sites, and then build the building. If we consider these quantities, the reuse of existing buildings can have significant energy benefits. Additionally, the National Trust for Historic Preservation reports that it can take 35 to 50 years to save the amount of energy lost when an older building is demolished (National Trust for Historic Preservation n.d.). But typically a greater amount of energy is used to operate an older building than a newer one. This can be due to lack of adequate insulation, leaky windows and doors, and inefficient lighting fixtures and heating and cooling systems. Therefore, existing buildings should be renovated when possible to make them more energy efficient. Even historic buildings can incorporate many energy-efficient upgrades without losing their character. The National Trust for Historic Preservation has a number of guidance documents available on this subject.

INCREASING THERMAL EFFICIENCY OF BUILDINGS

According to the American Institute of Architects (AIA), more than 75 percent of the buildings that will exist in 2035 will be either new or renovated (AIA n.d.). This presents an opportunity to make the built environment more energy efficient and climate friendly by greening the buildings that are projected to be built or renovated. Existing buildings can be made substantially more energy efficient through inexpensive practices that include taking advantage of site considerations, use of materials (such as window glazing and roof treatments), and the use of natural ventilation and daylighting. There are significant opportunities for planners to revise regulations, establish standards, and communicate best practices for significant energy savings and climate change mitigation purposes.

A number of techniques that can be used to improve or enhance the thermal efficiency of buildings are provided below.

Passive solar heating and cooling

Passive solar entails taking optimal advantage of solar exposure for both heating and cooling, as well as lighting. In all climates, buildings can benefit by keeping sunlight out during warmer weather, while in temperate and cold climates buildings can absorb, store, and release heat from their thermal mass in the winter. To achieve these gains, buildings must be well oriented to the sun's rays. This may require adjustments to subdivision and lot size standards. For example, minimum lot width requirements for new developments may need to be adjusted to allow for more lots that are optimally oriented for that climate.

Passive solar heating and cooling techniques depend upon climate conditions. What works in a temperate or cooler-climate setting may be ineffective or countereffective in hotter locations. The National Institute of Building Sciences's *Whole Building Design Guide* lists the following general passive solar design characteristics, but these characteristics must be tailored to particular climate conditions. (Note: "Skin-load dominated" buildings are typically residential or small commercial structures, while "internal-load" dominated buildings are generally larger commercial and industrial structures; the distinction involves the place in the structure where energy consumption primarily occurs.)

> Depending on climate, the passive solar design of skin-load dominated buildings might include:
> - orienting more windows to the south,
> - shading to avoid summer sun,
> - incorporating thermally massive construction materials,
> - providing properly sized and installed insulation and,
> - downsizing HVAC equipment.
>
> Depending on climate, the passive solar design of internal-load dominated buildings might include:
> - daylighting work spaces with properly oriented and controlled windows,
> - specifying high-performance glazing that reduce heat gain while admitting visible light,
> - selecting high-efficiency HVAC systems and,
> - incorporating adequate shading devices. (Fosdick 2008)

Consequently, in promoting energy efficiency through passive solar design, planners will need to ensure that zoning and subdivision regulations permit optimal structure orientations and that design guidelines and standards permit roof overhangs, operable windows, and skylights, as appropriate for climate conditions.

General Structure Design

There are a number of general structure design issues of interest to planners, such as building orientation, fenestration, and roof type, since they are potentially affected by design guidelines and standards, subdivision regulations, historic preservation requirements, lot size and lot width standards, and others. Planners should be aware of energy-efficiency issues involving structure design that may be affected by both general and specific code requirements and practices.

Windows

The Whole Building Design Guide notes that in "skin-load dominated structures (such as housing) optimum window design and glazing specification can reduce energy consumption from 10–50 percent below accepted practice in most climates" and that in "internal-load dominated commercial, industrial, and institutional buildings, properly specified fenestration systems have the potential to reduce lighting and HVAC costs 10%–40%" (Ander 2008). Therefore, window design and installation present planners with an important opportunity to achieve greater energy efficiency in both new construction and renovation. The primary issues for planners are to ensure that design guidelines and standards support energy-efficient window placement, as well as design and glazing alternatives, and to make sure that the public is fully aware of these potential benefits.

Roofs

In most U.S. climates, so-called cool roof design can significantly improve the energy efficiency of both newly constructed and renovated structures. The Lawrence Berkeley National Laboratory's Environmental Energy Technologies Division (EETD) has described the benefits, aesthetic challenges, and potential solutions to implementing this climate change mitigation technique:

> "Cool" roofs reflect more of the sun's radiation than do conventional roofs, lowering temperatures inside buildings, decreasing air-conditioning energy use, and reducing the "urban heat island." . . .
>
> Traditional cool roofs are white because light surfaces absorb less solar radiation than dark ones. . . . [R]aising the solar reflectance of a roof from about 20 percent (dark gray) to about 55 percent (weathered white) can reduce a building's cooling energy use by 20 percent. [However,] non-white cool roofs can be manufactured using colorants (pigments) that reflect the invisible, "near-infrared" radiation that accounts for more than half of the energy in sunlight. "Our research estimates that the potential net energy savings in the U.S. achievable by applying white roofs to commercial buildings and cool colored roofs to houses is valued at more than $750 million per year," says Hashem Akbari, head of the Heat Island Group at Berkeley Lab. . . .
>
> [W]idespread regional application of cool roofs can reduce ambient air temperatures and retard smog formation. Cool roofs can also reduce peak electricity demand in summer, which helps reduce strain on the aging electricity grid when relief is most needed. The lower temperatures of cool roofs may also increase the roofs' serviceable lives. . . .
>
> [T]he roofing industry has adopted voluntary standards for measuring the solar reflectance of roofing materials and has set up the Cool Roof Rating Council to develop labels that inform buyers about the relative degree to which various roofing products reflect solar radiation and emit heat through thermal radiation. The building materials industry has also introduced a number of products that help increase roof reflectance, mainly elastomeric coatings, single-ply membranes, tiles, and metal roofing. The ENERGY STAR® program certifies cool roof products with its voluntary label and offers a web-based guide to ENERGY STAR roof products. (Chen 2004)

Figure 10.1. Installation of light-colored roofing in order to better reflect sunlight and reduce interior temperature

As with windows, planners need to examine their regulations and guidelines to ensure that cool roof technology can be used in their communities and that the public is aware of the efficiency potential inherent in the technology.

Green Roofs and Green Walls

Green roofs and green walls introduce living plant material to the exterior of structures. Such features reduce building heating and cooling needs, sequester carbon, filter other air pollutants, and help to manage stormwater. They also:

- Protect underlying roof material by eliminating exposure to the sun's ultraviolet (UV) radiation and extreme daily temperature fluctuations.
- Serve as living environments that provide habitats for birds and other small animals.
- Offer an attractive urban amenity for residents or building tenants.
- Reduce noise from the outdoors.
- Reduce urban heat island effects.

Green roofs are typically constructed of a lightweight engineered soil media, underlain by a drainage layer, and a high-quality impermeable membrane that protects the building structure. Green roofs help to mitigate climate change by reducing a building's cooling needs in the summer and heating needs in

Figure 10.2. The roof of the 12-story Chicago City Hall has been retrofitted with a 22,000-square-foot rooftop garden. The primary goal of this installation, which was completed in 2001, was to demonstrate that green roofs help to reduce urban air temperature.

the winter. They also increase the albedo, or reflectiveness, of the roof over traditional roof design, allowing energy to be reflected rather than being stored and contributing to the urban heat island effect. Green roofs have also been shown to lengthen the life of underlying roofing materials, saving money over the life of the building, and to mitigate the effects of urbanization on water quality by filtering, absorbing, or detaining rainfall.

STRUCTURE AND SITE HAZARDS RESULTING FROM CLIMATE CHANGE

There are five major climate change hazards for structures and sites: hurricane winds, storm surge, precipitation, wildfires, and permafrost thaw. These hazards present regional threats to buildings and sites that can sometimes be addressed through design techniques.

Hurricane Winds

Hurricanes are projected to become stronger as the climate warms. As a result, structure design in vulnerable areas will need to account for higher winds.

Hurricane winds not only can damage site features like landscaping but can make some kinds of them into airborne objects that damage structures. Careful landscaping species selection and placement, retention of native vegetation, and placement of structures on the least vulnerable areas of sites can minimize potential for damage.

Figure 10.3. Wind damage

Storm Surge

Stronger tropical systems, combined with rising sea levels (and subsidence in certain areas), can exacerbate the potential for damage from storm surge. As with hurricane winds, structure design must account for such eventualities. Emergency management planners have tools available to them like the SLOSH (Sea, Lake, and Overland Surges from Hurricanes) model to evaluate the potential extent of storm-surge heights and winds. Typically used for evacuation purposes, the SLOSH model may also be useful to planners in determining building setback and first-floor structure elevation in accounting for storm-surge effects (National Hurricane Center n.d.).

Storm surge can create significant damage to site features like landscaping during the surge event. Postsurge damage to landscaping can occur through the introduction of salinity to the soil or groundwater. Selection of salt-tolerant plant species is one technique that can be used to minimize site damage associated with storm surge.

Figure 10.4. Storm surge from Hurricane Dennis, 2005

Heavier Precipitation

Climate change is projected to create a greater number of heavy precipitation events, resulting in the potential for greater flooding, erosion, landslides, mudslides, avalanches, snow-load damage, and other precipitation-related impacts that can affect structures. Building codes and development standards should take these types of events into account in order to adapt to climate change effects. For example, flood elevation maps should be reevaluated and snow-load code standards in cooler climates should be reconsidered.

Heavier precipitation has the potential to damage site features through erosion and inundation. Rapidly moving stormwater can erode parking lots and site landscaping. Flooding can cover plants with water or sediment, smothering root systems. Careful management of stormwater flow across properties and selection of flood-tolerant landscaping can reduce damage potential from heavier precipitation events.

Wildfires

Drought conditions created by climate change can increase the potential for wildfires to occur. As a consequence, structure design should account for this potential threat in areas that are vulnerable to wildfires.

One national program aimed at addressing wildfire safety is Firewise Communities, which aims to involve "homeowners, community leaders, planners, developers, and others in the effort to protect people, property, and natural resources from the risk of wildland fire—before a fire starts. The Firewise Communities approach emphasizes community responsibility for planning in the design of a safe community as well as effective emergency response, and individual responsibility for safer home construction and design, landscaping, and maintenance" (see www.firewise.org). The Firewise guidelines for structure design include:

> **Roof:** The roof can be the most vulnerable to burning embers during extreme wildfires. Installing fire-resistant roofing material with a Class A, B, or C rating, such as composition shingle, metal, and clay or cement tile, will help keep flames from spreading.
>
> **Walls:** Materials that resist heat and flames include cement, plaster, stucco, and masonry, such as concrete, stone, brick, or block. If your home has vinyl siding, use metal screening over openings that may become exposed if the siding were to melt due to heating during the wildfire.
>
> **Windows:** The heat of a wildfire can cause glass on exterior windows to fracture and collapse. Without a metal screen, a collapsed window will allow firebrands to enter and ignite the house. Double-paned glass can help reduce this risk. Tempered glass is the least likely to break due to the heat of a wildfire. For skylights, go with glass; it withstands higher temperatures than plastic or fiberglass.
>
> **Openings/Attachments:** Eaves, fascias, soffits, and vents should be "boxed" or enclosed with metal screens to prevent objects larger than ⅛" from entering the home. The undersides of overhangs, decks, and balconies should be screened or enclosed with fire-resistant materials. Make sure fences constructed of flammable materials, such as wood, don't attach directly to your home—if it's attached to the house consider it part of the house. (Firewise Communities n.d.)

Firewise further notes that wildfires create specific site hazards, especially within what is known as "the home ignition zone," which "consists of the home and its immediate surroundings within 100 to 200 feet." Removal of understory plantings, elimination of potential fuel materials such as fallen leaves and brush, use of nonflammable material for fences, and "limbing up" mature trees are techniques which can be used to reduce site vulnerability in that zone (Firewise Communities n.d.). For more information see U.S. Forest Service n.d. and Schwab and Meck 2005.

Permafrost Thaw

Areas that have permafrost are particularly sensitive to climate warming. According to the Geological Survey of Canda (GSC), "Much of permafrost is at temperatures very close to the melting point of ice, and processes triggered by thawing ground ice are sensitive to warming, particularly where ice contents are high and the potential for soil instability upon thaw exists" (GSC n.d.). As a result, structures built on permafrost may become unstable as permafrost thaws. This can result in structural damage as an unstable foundation shifts and exposes walls to new stresses.

Permafrost thawing can create a variety of site impacts, including unstable parking lots and facilities, broken utility lines, and hazards from landslides. Construction techniques that minimize damage to site features should be followed for new construction and renovations. Location of structures on the least vulnerable areas of sites, such as away from steeper slopes, can reduce damage potential from landslides.

In the United States, this problem exists primarily in Alaska.

Figure 10.5: A permafrost thaw–induced landslide in Canada

LANDSCAPING AND SITE DESIGN ISSUES

Site design is a critical component in mitigating climate change through increased energy efficiency. Site landscaping shades individual structures and reduces the urban heat island effect for surrounding areas; green infrastructure achieves those same purposes while also reducing the physical cost of stormwater management. Several specific site design techniques are discussed below.

Water Management

Drought and heavier precipitation events create challenges for site landscaping. Solutions can include selection of proper landscape plants, low-volume or alternative irrigation techniques, xeriscaping, or tree protection construction techniques.

Plant selection. The use of native species is generally recommended in most adaptive landscaping applications for two reasons:

- Native plant varieties generally have deeper root systems than exotics. This makes them more adept at capturing and filtering rainwater and surviving drought, and it gives them a higher capacity to sequester carbon.
- In general, native landscaping requires less maintenance in the forms of mowing, fertilization, and pesticide application. This eliminates the need to use combustion-powered equipment, reduces fuel used to transport materials and equipment, and eliminates chemicals and fertilizers that often require additional irrigation.

These considerations may not always apply, since warmer temperatures and changing precipitation patterns may affect the viability of certain current native species. Planners can develop lists of climate-resilient species for landscaping guidelines or standards that reflect their particular climate.

Low-volume or alternative irrigation sources. According to the Xeriscape Council of New Mexico (www.xeriscapenm.com), as much as 70 percent of water in a municipal water system can go to residential use, and of that residential use, almost half can be used to maintain landscape. Reducing the volume of water necessary to maintain landscaping is a major area for action for most water systems.

Low-volume irrigation techniques, equipment and devices are specifically designed to limit the volume of water applied and efficiently deliver that water within the root zone of the plant. Examples of low-volume irrigation techniques and equipment that might be used at the site level include hand watering and the use of soaker hoses and microirrigation equipment, such as emitters and drip tubes (Southwest Florida Water Management District 2003).

Rather than from wells or public water systems, irrigation water supplies can come from *alternative sources*, including graywater, reclaimed water, and rain barrels and cisterns.

- Graywater from bathtubs, showers, and clothes-washing sources can be collected and stored to be used for irrigating lawns and other site landscaping if permitted by local building and health codes. Other wastewater cannot be directly used for this purpose, due to public health concerns.

- Reclaimed water (sometimes called reused water) uses treated domestic wastewater for irrigation purposes. Unlike graywater, this wastewater is piped to and treated at a sewage treatment facility and then piped back for irrigation purposes, using a separate distribution system usually distinguished from potable-water distribution systems by the use of purple-colored pipes. (Hence, such systems are sometimes called "purple pipe" systems.)

- Rain barrels and cisterns are low-cost, effective, and easily maintainable detention devices that have residential and commercial applications. Rooftop runoff is channeled into rain barrels and cisterns where it can then be used for landscape irrigation.

Figure 10.6: Typical rain barrel setup for a residential property

Xeriscaping. A xeriscape is a landscape that uses little supplemental water. It does not necessarily refer to a dry, barren landscape but to one that is designed with water conservation in mind. A xeriscape may be designed to also minimize labor, energy, and chemical inputs over time.

Some cities, such as Albuquerque, New Mexico, and Las Vegas, Nevada, are using xeriscapes on public lands. They have also created incentive programs to pay residents on a square foot basis to transform existing water-intensive lawns to xeriscapes. Such retrofit programs can have significant effects on overall water use for irrigation purposes.

Benefits of Large Tree Retention

While there is some debate over whether large-scale tree planting is an effective means of carbon storage in northern latitudes (see Gibbard et al. 2005), in urban or developed areas tree planting has a cooling effect due to the relatively lower albedo of parking lots and rooftops. Shading buildings and parking lots can reduce cooling loads for buildings by reducing the ambient air temperature. The amount of reduction varies by location, climate, amount of paving and rooftops, type of vegetation, and other factors, but it can be significant, particularly in hotter climates.

Sources of information about the benefits of urban trees include:

- The U.S. Forest Service Center for Urban Forest Research (CUFR; see www.fs.fed.us/psw/programs/cufr), which provides a large number of urban forest studies in different areas of the country for comparison purposes. CUFR also provides i-Tree, a computer program that allows communities to assess the benefits of their urban forests and street trees.

- The EPA heat island Mitigation Impact Screening Tool (MIST; see www.epa.gov/hiri/resources/tools.html#MIST), which enables communities to gauge the effectiveness of increasing albedo and enhancing the amount of vegetation.

Consequently, site landscaping, particularly when it results in the retention, planting, and maintenance of larger trees, has the potential to significantly reduce energy usage, thereby helping mitigate the effects of climate change.

Tree Protection Construction Techniques
Site design should include not only the placement of new trees but also the protection of existing trees. Existing trees do not require the irrigation and care young trees need. Additionally, existing trees are generally acclimatized to flooding conditions in areas subject to periodic flooding.

Green Infrastructure
Green infrastructure can include site-specific design features like swales and rain gardens, but it can also be more comprehensive in scope. For example, it could entail an interconnected network of open spaces and natural areas that might include greenways, wetlands, parks, conservation and preservation areas, and flood-prone areas. Green infrastructure benefits can include improving a community's ability to manage stormwater by controlling flooding and improving water quality, maintaining wildlife habitat and travel corridors, supporting or providing recreational opportunities, and preserving and creating open space. With regard to climate change mitigation, green infrastructure is primarily important for its role in preserving native vegetation that sequesters carbon dioxide and in reducing the urban heat island effect by maintaining tree canopies, streams, and other water bodies, resulting in cooler temperatures.

Permeable Pavement
Permeable pavement allows stormwater to penetrate pavement surface so it can be channeled into retention areas or absorbed by underlying soil. It is also used as an urban heat island mitigation tool. In some climates, permeable pavement is less durable than standard pavements, so it requires particular maintenance.

OPTIONS FOR LIGHTING AND FIXTURES
Site Lighting
Minimizing the extent of site lighting is one way to reduce overall energy usage. Site lighting is installed for a variety of reasons. Insurance risk-management standards require a certain amount of illumination for safety purposes; driveways and walkways require illumination for both orientation and safety; and site designers often illuminate buildings and landscaping to improve aesthetics.

Using photovoltaic technology to collect and store solar energy during daytime hours for nighttime illumination can reduce energy usage for site lighting. Imagine the solar-energy collection potential of a canopy above fuel pump islands at a gasoline station or convenience store, which require significant site lighting for safety purposes. Using electronically managed lighting systems and energy-efficient lamps can also reduce the amount of energy used for site lighting. LED lights use significantly less energy than other bulbs and last much longer.

Planners can promote awareness programs and require submission of site lighting plans to encourage conscious decision making about energy usage. Some communities may wish to pursue "dark skies" initiatives to minimize overall light pollution; site lighting plans for individual developments may then be required to conform to those standards.

Plumbing Fixtures and Water Use

Since drought conditions may be exacerbated by climate change, reducing water demand is an important adaptation technique. Reduction in water consumption also translates into a reduction in energy used to treat and deliver the water, which is a climate change mitigation benefit.

There are a variety of technologies for reducing water use in buildings that can be applied in both existing and new construction. Examples include the use of low-flow toilets, waterless urinals, graywater systems for toilet flushing, and low-flow showerheads; thoughtful placement of water heaters near points of use to minimize waste while waiting for hot water to reach a faucet; and the use of tankless or on-demand hot water heaters.

Most building codes contain provisions for low-flow toilets. In areas where water supply is problematic, these codes should be examined to ensure that other water-saving alternatives are permitted. Plan review and inspections programs can promote the use of water-saving technologies at the design level. Additionally, opportunities to retrofit older technology can be pursued through incentives or requirements, as in Santa Fe:

> Because pre-1994 toilets use two to three times more water than low-flow toilets, and the average adult flushes a toilet five to seven times per day, an old toilet can waste 23.8 gallons (90 liters) per person per day. With average indoor domestic use today at nearly 70 gallons (265 liters) per person per day, retrofitting three or four toilets would provide enough salvaged water for one person's daily use. Clearly, incentivizing retrofits can produce big water savings.
>
> Using this idea, Santa Fe, New Mexico, developed a toilet retrofit program so that the city's water utility would experience no new loss to the system despite growth. This program requires developers to retrofit existing toilets in existing homes before pulling a building permit. By doing so, the city's utility accumulates enough water savings to meet the demand created by each newly built home. (Henrie 2007)

"RENEWABLE READY" SITES AND BUILDINGS

Making buildings and sites "renewable ready" involves identifying, creating, and preserving opportunities to install site-level renewable energy and efficiency technologies—such as appropriate locations for small wind turbines, solar panels, biomass-fired district heating, and geothermal heating and cooling. Site features that support energy-efficient transportation choices, such as bicycle and pedestrian paths, bike racks, and well-designed transit stops, are also important.

Declining fossil fuel reserves, concerns about pollution and climate change, and a desire for energy independence are driving the decentralization of energy production in the United States. Wind, solar, and geothermal energy applications are now available and economically feasible at even the single-family home level. Microhydropower and cogeneration technologies are also available. Planners need to be aware of these applications and, where appropriate, encourage their incorporation into building design through ordinances that permit their use, regulatory incentives that encourage their use, and public relations initiatives that promote awareness of their effectiveness in local situations.

ANN ARBOR ENERGY CHALLENGE: A COMPREHENSIVE APPROACH FOR ENERGY WITHIN THE COMMUNITY

Population: 114,386

The City of Ann Arbor, Michigan, has been recognized as a leader in environmental quality nationwide. As a part of its comprehensive environmental policy, Ann Arbor has long been committed to raising awareness about energy issues. The city adopted its first energy plan in 1981 and updated it in 1994.

A new impetus was given to the city's environmental program in 2006 by "Ann Arbor's Energy Challenge." The challenge, which was passed as a city council resolution in May 2006, calls for Ann Arbor to use 30 percent green energy in municipal operations by 2010 and for the whole city to use 20 percent green energy by 2015. Conservation is stressed first, followed by an increased use of local renewable energy as well as a greater percentage of green fuel purchase.

In support of the challenge, the city has set up an action plan to boost the use of solar energy. In 2007, the city partnered with the U.S. Department of Energy to become designated as a Solar America City. This partnership enabled the city to receive federal grants and technical assistance for the implementation of a two-year $632,000 project to integrate solar energy throughout the community. An online tool is available on the city's website that provides any Ann Arbor resident with an assessment of the solar energy potential of his or her home. Several photovoltaic and thermal solar panels will be installed on five city facilities. Outreach and educational efforts are also a major component of the plan, with city-organized solar energy workshops, teacher training on solar energy curriculums, and aggregate solar technology purchasing programs for residents, which reduce the initial cost of solar installations through discounted group rates.

COLORADO SOLAR READY HOMES BILL

In May 2009, Colorado governor Bill Ritter signed into law House Bill 1149, "The Solar Ready Homes Bill," a measure that will encourage construction of "solar ready homes" as part of the state's long-term energy strategy. The law requires that builders of newly constructed detached single-family residences (1) offer prospective home owners the option of having their home built "solar ready" and (2) provide home owners with a list of qualified contractors who can determine and install the most appropriate solar energy technologies for a particular situation. This policy encouraging the incorporation of renewable design features into new buildings is the first of its kind in the United States.

The law is intended to give Coloradans access to affordable clean energy. Analysis has shown that a significant percentage of the cost of a solar energy project can be mitigated by incorporating renewable energy features in the construction process (DOE 2007b). A provision of the law that allows home buyers to incorporate solar technology purchases into their home mortgages helps alleviate concerns around high initial costs, which can be a barrier for individuals purchasing solar technologies (Ritter 2009).

The law requires home builders to offer home buyers one or more of the following:

- A residential photovoltaic solar generation system or a residential solar thermal system
- Upgrades of wiring and/or plumbing planned by the builder to accommodate future installation of such systems
- A chase or conduit constructed to allow ease of future installation of the necessary wiring or plumbing for such systems.

The law continues the work the State of Colorado has been doing related to energy and climate change issues. The Colorado Climate Action Plan, released in 2007, targets both adaptation and mitigation of climate change through coordinated policies, incentives, and regulations. State leaders have also sought to create a "New Energy Economy" by harnessing opportunities associated with reducing emissions. The advancement of low-carbon businesses and new energy jobs are central to the state's plan for a new energy future. The Solar Ready Homes Bill has the potential to inspire innovation in renewable energy technology research and development and help bring both economic and climate sustainability benefits to the state.

Planners must also address regulatory roadblocks to the adoption of alternative energy systems, including failures to address solar access in local regulations; design guidelines or aesthetic standards that preclude installation of solar collectors; height limits that preclude wind turbines; lot clearance or grading limits that prohibit the installation of geothermal pipe fields; and failures to account for cogeneration in local zoning regulations. Planners need to examine their code requirements to accommodate site-level alternative energy options in ways that are balanced with other community objectives.

Solar

Opportunities for installing solar energy capacity vary from site to site and building to building. Still, evaluating the solar potential of any given site or building and identifying where solar panels, passive water heating, or other solar technologies might go—and protecting that solar potential during the planning and development process—can help promote deployment of solar energy over time. (See the Colorado case study.)

Wind

Similarly, the wind potential of sites and individual buildings also should be evaluated. Wind potential can be particularly variable. Wind installations on buildings typically require structural reinforcements beyond normal building framing.

The American Wind Energy Association (www.awea.org) provides information about wind turbines in various applications, including at the single-family residential level.

Biomass

Accommodating biomass heating at the site and building scale involves installing a suitable furnace and providing for convenient means for supplying and storing the fuel. Installing and fitting a biomass furnace is generally the same as doing so for a typical furnace. Supply and storage of biomass fuel is different, however, than the infrastructure that may accompany a natural gas, propane, or heating oil system. Wood chips, pellets, or other forms of biomass fuel are solid and are typically delivered as a loose pile or in large bags. Planners can make regulatory provisions for the accommodation of protected and convenient areas for fuel delivery and storage, which will enhance the feasibility of installing biomass energy.

Geothermal Heating and Cooling

For on-site uses of geothermal heating and cooling, it is important to designate suitable areas for installing pipes that will circulate through deeper earth layers and to approximate how such pipes will connect to a building. For off-site potential, anticipating and providing proper access for hookups to a district system, usually routed with other sewer or water piping, is the most important consideration.

Electric Systems

Distributed renewable energy, use of electricity to fuel personal vehicles, and advanced systems to manage energy use will benefit from planners' attention to what connections may need to be added to electric systems. How and where control panels and monitoring instruments, fuel cells, charging outlets, and connections to nearby renewable electricity generation will be located and routed will help ease the transition to a smarter and more efficient electric system and increase opportunities to install advanced and renewable energy technologies.

SUMMARY

Due to the extent to which buildings consume energy, especially electricity, energy-efficient structure design plays an extremely important role in mitigating climate change. Site design has the potential to enhance the energy efficiency of structures, both on-site and off-site. Additionally, structures and sites have location-specific vulnerabilities to climate change that must be identified and addressed through effective adaptation responses, includ-

TOOLS FOR GUIDING BUILDING AND SITE DESIGN DECISIONS

Visioning and Goal Setting	Set goals for reuse of existing buildings, retrofits, and renewable-ready sites
	Incorporate long-term energy efficiency, conservation, and climate resilience in community goals and objectives
Plan Making	Identify possible structure and site hazards from climate change in local plans
	Identify impediments and challenges to implementing green development practices when developing plans
Standards, Policies, and Incentives	Create standards, policies, and incentives to promote energy-efficient, climate-resilient, and renewable-ready sites and buildings
Development Work	Encourage energy-efficient and climate-friendly building and site design in new projects
Public Investment	Promote public investments in buildings and facilities that are energy-efficient and avoid or mitigate development in high-risk hazard areas, such as those prone to flooding or wildfires

CHAPTER 11

Natural Resources

 Natural resource management has an important role to play in both mitigating and adapting to the effects of climate change. For mitigation purposes, natural areas allow for carbon sequestration, helping to reduce the volume of GHG emissions that are released into the atmosphere. Natural areas can also help with climate change adaptation. For example, floodplains and other natural features can help buffer against flooding, storm surge, and other impacts of climate change.

Climate change is expected to modify natural systems, in many cases significantly. For example, rising sea levels are projected to alter the salinity of low-lying coastal marshes, drought is expected to affect the habitat of many plant and animal species, and rising temperatures may extend the ranges of some species while shifting or contracting those of others. These changes could affect food supply, species diversity, timber harvests, and many other important components of human relationships to the natural world. As a consequence, management of natural resources will become increasingly important as the effects of climate change materialize and ecosystems react and are modified.

This chapter explores the ecosystem impacts that may result from climate change, how planners can help mitigate these impacts, and how communities can adapt to local and regional changes. The following specific areas are discussed:

- Carbon Sequestration
- Water Management
- Agriculture
- Forestry and Fire Management
- Ecosystem Management
- Additional Planning Tools for Natural Resource Management

CARBON SEQUESTRATION

Natural systems sequester carbon, slowing or inhibiting its concentration in the atmosphere. According to a 2009 report by the U.S. Environmental Protection Agency (EPA 2009b), U.S. net carbon dioxide emissions would be approximately 15 percent (or one billion metric tons) higher if not for sequestration by forests and grasslands. Thus, maintenance and enhancement of natural sequestration systems are highly important factors in climate change mitigation.

Sequestration can take many forms, including highly technological "fixes," such as injecting carbon dioxide deep into the oceans where it is sequestered by water pressure or injecting it underground in an extractive process that displaces methane for energy uses while sequestering the CO_2. For the most part, however, technological approaches are not the province of local planners since they involve large-scale private or federal and state government initiatives. Rather, planners can promote natural sequestration through such activities as preserving forests and open space, conserving rural land uses, conserving parkland (both urban and rural), creating urban forestry programs, and managing open land to maximize carbon sequestration.

Mitigation of climate change through natural resource management literally and figuratively covers a lot of ground, including agriculture and forestry practices, rural land-use patterns, conservation practices and incentives,

TABLE 11.1. CARBON SEQUESTRATION RATES FOR SELECTED AGRICULTURAL AND FORESTRY ACTIVITIES

Activity	Representative rate in U.S. (metric tons of carbon per acre per year)
Afforestation	0.6–2.6
Reforestation	0.3–2.1
Changes in forest management	0.2–0.8
Conservation or riparian buffers	0.1–0.3
Conversion from conventional to reduced tillage	0.2–0.3
Changes in grazing land management	0.02–0.5
Biofuel substitutes for fossil fuels	1.3–1.5

Source: Adapted from www.epa.gov/sequestration/rates.html

KING COUNTY, WASHINGTON
Population: 1,875,519

Natural resource management offers a significant opportunity for communities to enhance their climate change resilience. King County, Washington, which encompasses Seattle and the Puget Sound region, has taken a leadership role in incorporating natural resource management into its adaptation planning. The county has worked closely with the University of Washington Climate Impacts Group, an interdisciplinary research team studying the impacts of global climate change on the Pacific Northwest, to translate scientific climatic data into policies and projects relevant to King County's natural resources.

Natural resources are affected over time by many factors. Historically, changes in land-use patterns, new development in floodplain areas, natural processes, and changes in water flow management have had the largest impacts on our most vital natural resources. Planners and community leaders must be prepared to mitigate and adapt to climate change impacts on our ecosystems.

King County is characterized by its abundant natural resources, including the Cascade Mountains, extensive river systems, glaciers, and forests. One of the largest projected impacts on King County from climate change is a reduction in the extent of the Cascade Mountains snowpack, which acts as a natural reservoir for water storage and is the main supplier of hydropower and drinking water for the county. The snowpack has already decreased significantly over the last several decades due to warming during the 20th century. Predicted increases in temperatures and changes in precipitation associated with climate change may cause further reductions, resulting in higher winter stream flows and lower summer stream flows. Such alterations could affect water withdrawal and recharge, resulting in a decline in predictable yields from the region's water-supply reservoirs (King County 2007).

Among the adaptive measures King County has taken to prepare for climate change impacts on water infrastructure are improved flood planning, more than $300 million of improvements to levees, and an increase in the amount of water reclaimed from wastewater treatment plants. King County has also made investments to adapt to anticipated natural resource management needs driven by climate change. In 2006, King County executive Ron Sims issued executive orders requiring the county to make changes in a number of areas to help reduce greenhouse gas emissions and prepare for climate change impacts. The county chose to focus its efforts on environmental management—along with land use, transportation, and renewable energy policy—and to try to mobilize the community around common goals. For example, the King County Department of Natural Resources and Parks and the Water and Land Resources Division are working to determine how specific bodies of water in King County's river system will be affected by climate change and what adaptive measures are needed.

The county has produced several planning documents that are relevant in this context. The 2006 King County Flood Hazard Management Plan, the 2007 King County Climate Plan, and the 2008 update to the King County Comprehensive Plan

Photos by Ned Ahrens

Figure 11.1. (Above) Cedar River near Madsen Creek, flooded in January 2009

Figure 11.2. (Left) Snoqualmie Falls on the Snoqualmie River, flooded in January 2009

address climate and its impacts on natural resources, including the management of watershed and water resources, carbon sequestration, and ecosystem resiliency.

The King County Comprehensive Plan incorporates climate-related mitigation and adaptation goals into the long-term vision and goals for the region. Chapter 3 of the 2008 plan update integrates climate change assessment, mitigation, and adaptation into rural communities and natural resource lands planning. Climate change–related policies include:

- Recognizing the role of resource lands in supporting carbon sequestration.

- Describing potential impacts of climate change impacts on forests and recognizing the climate change adaptation benefits of long-term forest management.

- Recommending the county consider climate change impacts on forestry and take steps to improve forest health and resiliency through technical assistance, management of county-owned lands, and support of neighborhood-based efforts to reduce risk of wildfire.

(continued on page 136)

(continued from page 135)

▶ Recommending that the county collaborate with other entities to assess the likely impacts of climate change on agriculture and to develop mitigation and adaptation strategies suited to King County soils and farm economy, and also recommending that this information be made available through technical assistance and farm planning programs (King County 2008).

King County has also been working to increase the carbon sequestration capacity and climate change resiliency of private and public lands (King County n.d.). The county is waiting for final approval of a proposal, in partnership with the National Wildlife Federation, to develop and implement an Urban and Community Forestry Climate Preparedness and Response Program. The program would provide urban and suburban property owners with guidance and incentives to sequester carbon through forest management activities. Relevant actions include expanding protected areas, diversifying tree species, removing invasive species, and otherwise removing carbon dioxide from the atmosphere and storing it in biomass. King County's program will provide tools and incentives that can serve as models for other communities seeking to maximize the mitigative and adaptive values of urban and community forests.

King County believes that policies, planning, education, programs, partnerships, advocacy, service delivery, and infrastructure should all be viewed through a climate change lens. The county continues to focus on long-term climate change mitigation and adaptation efforts and to make new plans for preparing and responding to climate changes. ◀

and tax policies. In general, mitigation of climate change through the use of natural resource areas and ecosystems involves establishing or maintaining activities that promote carbon sequestration. Table 11.1 (above) provides information about the relative sequestration rates for a variety of activities.

Agriculture and Forestry Practices

Farmland and forests have a high potential for carbon sequestration. Agricultural crops sequester carbon by turning it into plant material and by storing carbon in soil. Many sustainable agricultural practices significantly increase the amount of carbon that can be sequestered in soils. These include planting cover crops, using no-till farming techniques, adding organic material (crop residues, biosolids, compost) to soil, planting more deep-rooted perennial crops, and limiting the use of chemical fertilizers, which disrupt natural soil processes.

Research shows that organically managed soils can sequester more than 1,000 pounds of carbon per acre. In 2005, the 2.4 million acres of organically managed farmland in the United States (0.5 percent of all U.S. cropland) captured an estimated 2.4 billion pounds of atmospheric carbon (Rodale Institute n.d.). Planners can support organic soil management by promoting and allowing urban agriculture, local green foods movements, and similar activities.

Forestry operations also provide significant opportunities for sequestration, as illustrated in Table 11.2.

TABLE 11.2: FORESTRY PRACTICES THAT SEQUESTER OR PRESERVE CARBON

Key Forestry Practices	Definition and Examples	Effect on Greenhouse Gases
Afforestation	Tree planting on lands previously not in forestry (e.g., conversion of marginal cropland to trees).	Increases carbon storage through sequestration.
Reforestation	Tree planting on lands that in the more recent past were in forestry, excluding the planting of trees immediately after harvest (e.g., restoring trees on severely burned lands that will demonstrably not regenerate without intervention).	Increases carbon storage through sequestration.
Forest preservation or avoided deforestation	Protection of forests that are threatened by logging or clearing.	Avoids CO_2 emissions via conservation of existing carbon stocks.
Forest management	Modification to forestry practices that produce wood products to enhance sequestration over time (e.g., lengthening the harvest-regeneration cycle, adopting low-impact logging). Forest thinning also can reduce the risk of disease and forest fire.	Increases carbon storage by sequestration and may also avoid CO_2 emissions by altering management. May generate some N_2O emissions due to fertilization practices. Removing underbrush will decrease the amount of carbon stored. However, not only do wildfires destroy all of the carbon stored in trees and soils, but when the soot falls on snow or ice, it accelerates melting.

Source: Adapted from www.epa.gov/sequestration/forestry.html

Natural Areas

Natural ecosystems, farmland, and forests, when properly managed, have a huge potential for carbon sequestration. These areas are carbon sinks, storing carbon for long periods of time in plant material and soils. When these areas are developed, they not only lose the ability to continue sequestering carbon but also release stored carbon when organic soils are disturbed or when trees and vegetation are burned or decompose.

It is important to understand the sequestration capacity of ecosystems in a region and to protect it, since "enhancing the natural processes that remove CO_2 from the atmosphere is thought to be the most cost-effective means of reducing atmospheric levels of CO_2" (U.S. Department of Energy n.d.). There are a number of ways to promote conservation of natural resources, agricultural areas, and forest areas, including conservation easements and use-value taxation, which are discussed below.

Urban Forests

Planting trees and other vegetation in urban areas can have a significant impact on GHG sequestration and on the reduction of energy use in buildings that are cooled by them. The Colorado Tree Coalition (www.coloradotrees.org) has summarized key points on this subject, including:

- Approximately 800 million tons of carbon is stored in U.S. urban forests.
- A single mature tree can absorb carbon dioxide at a rate of 48 pounds per year and release enough oxygen back into the atmosphere to support the respiration of two human beings.

WATER MANAGEMENT

Water resource management will represent one of the major challenges of climate change adaptation across the United States. Specific issues will vary by region and even locality, but maintaining an adequate quantity and quality of water will be a virtually universal problem. Stormwater management will also be a common problem as precipitation intensity increases or as more rapid snowpack melts occur. See Chapter 7 for more detail.

According to the U.S. Government Accountability Office, at least 36 states will face severe water shortages due to a combination of drought, rising temperatures, urban sprawl, and population growth (GAO 2003). In October 2007, Georgia's Lake Lanier (a 38,000-acre reservoir that supplies more than three million residents with water) was less than three months from depletion, causing the governor to declare a state of emergency. Scientists studying drought in the Southwest have proven (by looking at clues such as the rings of thousand-year-old trees) that the last century was unusually wet

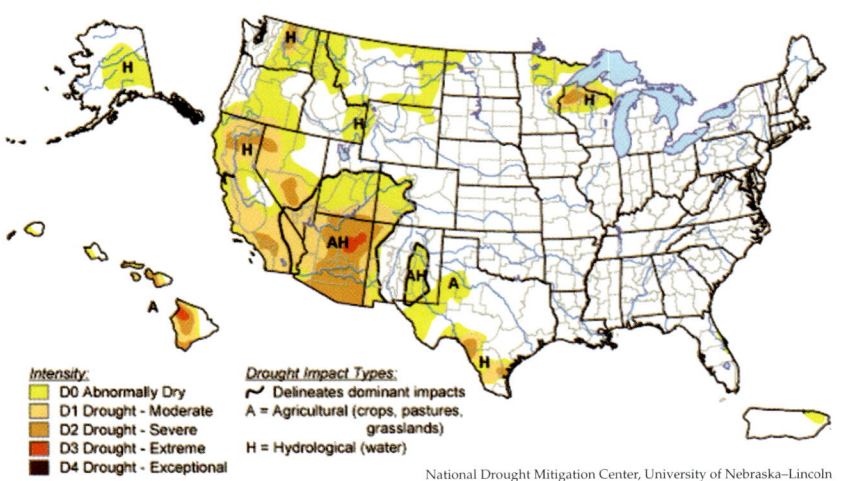

Figure 11.3. U.S. Drought Monitor

and that the region is now transitioning into more typical, drier conditions. Communities must change how they manage water resources if they are to continue to adapt and thrive. One source of information about drought issues is the U.S. Drought Portal (www.drought.gov), which regularly posts relevant news stories. Another source is the U.S. Drought Monitor (Figure 11.3), which is maintained by the National Drought Mitigation Center at the University of Nebraska–Lincoln.

Approaches that can be taken to manage water at the local level include:

- Green infrastructure
- Open space and water
- Native species; xeriscaping
- Water conservation
- Alternative water sources

Green Infrastructure

Green infrastructure can play an important role in mitigating climate change by sequestering carbon. Green infrastructure also can help communities adapt to the effects that climate change will have on water resources. As discussed in Chapter 3, climate change is anticipated to cause more severe variations in precipitation, resulting in more extreme storms and droughts. With more intense rain events come the threats of flooding and erosion, which damage property and cause pollution in waterways. Longer periods of drought shrink the existing water supplies that communities depend on. Green infrastructure provides a natural way to adapt to these changing conditions as discussed below.

Flood control. Increased impervious surface and the construction of stormwater drainage linking urban runoff with streams and rivers are two factors that greatly increase the "flashiness" of streams (i.e., their flooding is often more rapid, more severe, and more frequent than under natural conditions). Establishing parks, open space, or greenways in floodplains, wetlands, and in riparian zones allows bodies of water to behave naturally in response to large amounts of precipitation. Greenways established in floodplains provide inexpensive flood and erosion control by allowing a stream to naturally flood and deposit sediment; they are also places where excess water can naturally infiltrate the soil. Channelization of streams and development of their banks only increases flooding downstream.

Erosion control. Vegetated open space permits the infiltration of rain into the soil and helps slow the momentum of overland flow. Runoff that is slowed and dissipated is less able to erode stream banks and hillsides, decreasing erosion and sediment loading in streams.

Stormwater management. According to the EPA, "Runoff from urban areas is the largest source of water quality impairments to surveyed estuaries (areas near the coast where seawater mixes with freshwater)" (EPA 2008). Riparian buffers are an excellent way to filter and trap harmful pollutants and sediment, keeping them out of streams and rivers. Many plant species are especially adept at taking up excess pollutants and nutrients. These "hyperaccumulators" can be included in green infrastructure projects to tackle site-specific pollutants (such as nitrogen and phosphorous from golf course or homeowner lawn fertilizers).

Natural recharge. Pervious surfaces and open space allow water to infiltrate naturally to the soil, where some of it eventually travels to recharge surface and ground water. Prolonged droughts currently threaten water supplies all over the country; by promoting natural recharge, planners can help ensure continued availability of water resources.

Open Space and Water

Water resources in many parts of the country are extremely vulnerable to climate change. Development patterns and changing uses of these resources also play big roles in determining the availability of water for human consumption, industry, wildlife, and recreation. To protect water resources, planners and local governments need to implement practices aimed at:

- Protecting the quality of surface and groundwater by limiting pollution and sedimentation from runoff and erosion.
- Ensuring maximum recharge of ground and surface water sources.

Preserving and creating open space is one of the best ways that communities can protect their water resources. Trees and vegetation reduce topsoil erosion by breaking the fall of rainwater, slowing down water runoff. Pervious surfaces and vegetated areas allow rainwater to infiltrate soil slowly, rather than quickly running into a stream, ensuring that ground and surface water supplies are continually being replenished.

Vegetated areas act as natural pollution filters. Their canopies, trunks, roots, and associated soil and microorganisms filter particulate matter out of the flow toward storm sewers. Reducing the flow of stormwater reduces the amount of pollution that is washed into a drainage area. Plants take up nutrients such as nitrogen, phosphorus, and potassium, which pollute streams.

Native Species and Xeriscaping

As discussed in Chapter 10, using native plant species and xeriscaping help communities adapt to climate change by protecting water resources and requiring fewer fossil fuels for maintenance operations for the following reasons:

- *Native plants use less water*. Native plants are adapted to the local area's climate and require less water to survive, even as environmental conditions change.
- *Native plant varieties generally have deeper root systems than ornamental exotics.* This means that they are more adept at capturing and filtering rainwater, surviving drought, and sequestering carbon.
- *In general, native landscaping requires less maintenance in the form of mowing, fertilization, and pesticides.* This eliminates the need to use combustion-powered equipment, reduces fuel used to transport materials and equipment, and eliminates chemicals and fertilizers that pollute soil and water and harm beneficial insects and wildlife.

Water Conservation

There are many tools that communities can use to encourage water conservation. Restrictions on water usage can be effective as long as they equitably distribute costs among users and establish penalties that are effective. Some communities have banned new lawns and encourage residents to replace existing ones with xeriscaping. Also, many communities now offer rebate programs (either through local government or public utilities) for residents who replace old appliances with newer ones that use less water and energy. Making water conservation tools (such as rain barrels and low-flow shower heads) available at little or no cost is another way that communities can encourage water conservation.

Alternative Sources

Communities that are facing or already experiencing water shortages are looking to procure new sources of water. Cities, especially in the western United States, are buying irrigation water from farmers or buying ranches and farms to tap groundwater. Some are turning to water sources that were previously uneconomical, such as desalinating seawater and briny aquifers. El Paso, Texas, has tapped one such aquifer with a new desalination plant that opened in August 2007. Some states, like Georgia, are fighting legal battles with other states in an attempt to access more sources or to withdraw more water from existing sources.

There are many important factors to consider when searching for alternative water sources. Planning to access more water in drought years is not necessarily a viable option because the current climate trend in many areas is toward increasingly drier conditions; current resources may simply be unavailable in the future. As droughts become more frequent or of longer duration, reliance on short-term fixes is potentially risky. New access to aquifers, even large ones, may not provide long-term solutions to water woes because aquifers frequently take an extremely long time to recharge, often hundreds or even thousands of years.

The ways that we use water have serious effects on the natural environment: native species depend on sources of freshwater, and desalination operations generate concentrated brine that must be disposed of. In California, increased water diversion from the Sacramento–San Joaquin River Delta to drought-stricken farms and cities to the south is being blamed for the near collapse of the state's salmon fishery (Chea 2008). The collapse and new restrictions on salmon fishing may have a significant impact on the state's economy; according to the American Sportfishing Association, there are 2.4 million recreational anglers in California who generate as much as $4 billion in economic activity each year.

AGRICULTURE

Climate change adaptation for agricultural systems requires revisions to water use and pest management techniques, as well as adjustments to crop and livestock selection due to higher temperatures and longer growing seasons.

Water management for agricultural purposes involves changes to irrigation practices and how precipitation is handled. Water resources for irrigation purposes are likely to become scarcer in many regions of the country due both to drought-related supply issues and to competition with other users, such as municipalities that require potable water supplies. In some coastal areas, saltwater intrusion may further limit water supplies available for irrigation. As a result, agricultural operations will need to adjust to more drought-resistant practices and crops.

In many areas, water for irrigation has been subsidized in order to encourage agriculture. For these areas, adaptation may simply involve implementing more efficient irrigation practices. In other areas, more widespread and far-reaching adjustments may be necessary in order to maintain viable agricultural operations.

Management of precipitation for agriculture requires establishing greater resilience against both excess water (due to high-intensity rainfall events and flooding) and lack of water (due to extended drought periods). A key element in responding to both problems is an abundance of organic matter in agricultural soil, which stabilizes soil and improves its ability to retain water. The best methods to maintain soil organic matter involve low tillage and maintenance of permanent soil cover. These practices reduce impacts from flooding, erosion, drought, and fluctuations in temperature. Other techniques to adapt agricultural practices to climate change are organic agriculture, crop rotation, agroforestry, crop-livestock associations, and the use of hedges and vegetative buffer strips (Interdepartmental Working Group on Climate Change 2007).

Saltwater intrusion is a threat to coastal-area agricultural operations. A May 2008 tidal-surge flooding of cropland in the Delaware Bay area provoked the following long-term problems:

> Salt water inundation occurred where fields were flooded with sea water, brackish water, or tidal surge water from the Delaware Bay. Salt contaminated soils will have several effects on crops. The first is osmotic where high salt levels in the soil solution will draw water out of germinating seedlings and the roots of plants, causing desiccation. In less severe cases, elevated salt levels will make it more difficult for plants to take up water, thus increasing water stress and reducing growth. The second concern is the toxic effect of salt water constituents. Excess sodium is toxic to crop plants. In addition, chloride from salt water can be toxic to many crops.
>
> Soils that have had salt water leach into them will have high osmotic conditions (high dissolved solutes) and high levels of sodium. Levels of overall salts, sodium, and chloride will be reduced with leaching from rainfall, but this may take a considerable amount of time, depending on the amount of rainfall, soil type, water table, and the presence or absence of salt water intrusion in the ground water. On a sandy loam soil, salt levels may be reduced to tolerable levels within a year's period of time. On heavier soils and soils with high water tables, it may take several years for salt levels to drop to acceptable levels. In areas where salt water ponded for long periods of time, also expect effects to last for several years. Other problems include salt water mixing with ground water contaminating shallow wells and tidal overwash into irrigation ponds, contaminating irrigation water sources. (Johnson 2008)

Sea-level rise will introduce salinity to many coastal agricultural operations through inundation and will exacerbate the effects of storm surge from tropical and nontropical storm events, making events like the Delaware Bay saltwater flooding more likely. Leaching of salts that reach these coastal soils will not be possible in the event of inundation and may become more problematic after surge events due to higher water tables and other effects from sea-level rise.

Planners in agricultural areas will need to become familiar with irrigation, precipitation management, and pest management techniques that best match anticipated local conditions in order to help agricultural operations adjust to ensure continued agricultural sustainability. Additionally, threats from saltwater intrusion require evaluation to determine if agricultural operations can be maintained in vulnerable coastal environments.

FORESTRY AND FIRE MANAGEMENT

Commercial forest management operations are expected to face many of the same problems from climate change as agriculture will. New or larger numbers of pests, temperature-induced changes to tree species viability, and reduced yields resulting from drought could complicate forestry operations. Additionally, drought conditions may create more frequent and more severe wildfires, creating an additional threat that must be overcome through sound management practices. As with agriculture, local planners must familiarize themselves with pest and wildfire management techniques in order to assist forestry operations in their areas to achieve sustainability.

Wildfires

The effects of climate change coupled with natural climate cycles (higher temperatures, longer droughts, and shorter winters) can be seen in changes in natural ecosystems. In the southwestern United States, trees are dying due to lack of water and infestations of bark beetles that survive less severe winters. The dry conditions and abundance of dead wood in western forests fuel wildfires that are burning hotter and causing more damage than ever before. These factors are further compounded by no-burn policies in ecosystems that depend on fire for their continued survival. Frequent small fires burn close to the ground at low temperatures, clearing out fallen deadwood and underbrush. Fire suppression policies allow this fuel to build up, eventually resulting in fires that are more detrimental to the ecosystem and more difficult to contain.

When possible, communities should adopt and implement ecosystem-based management for wildfires. This kind of natural resource management "is a process that integrates biological, social and economic factors into a comprehensive strategy aimed at protecting and enhancing sustainability, diversity and productivity of our natural resources" (State of Michigan n.d.). Such a management program can be a component of a Community Wildfire Protection Plan (CWPP).

A CWPP is an important tool that planners can use to address wildfire problems. CWPPs were specifically provided for in the Healthy Forests Restoration Act (HFRA) that was enacted in 2003. The CWPP provisions in HFRA offer a new opportunity for communities in wildfire-prone areas to work closely with the U.S. Forest Service and the Bureau of Land Management to address wildfire issues in a comprehensive way. CWPPs can be used to increase awareness of the impacts of prolonged fire suppression, provide information about implementing wildfire protection measures both communitywide and on individual properties, manage expectations of property owners about the need for and benefits of prescribed burns as a management tool, and establish a productive dialogue among relevant federal agencies and the communities that they serve.

CWPPs are developed through an eight-step process (Communities Committee et al. 2004). A plan developed through this process can prove very valuable for local planners and emergency-services providers in addressing community concerns about wildfires as well as in helping mitigate the effects of wildfires, including their production of GHGs.

One component of these plans should be expectation management: persons living in heavily forested areas should be educated about the necessity of prescribed burning to reduce understory biomass to lower potential wildfire intensity. While there may be concerns about the aesthetic impacts of prescribed burns or mechanical removal of brush, property damage would likely be significantly more severe in the event of an actual wildfire without the use of these forest management tools.

A second component involves postfire vegetation management. Whether the fire event was a prescribed burn or a wildfire, steps should be taken to eliminate nonnative, invasive vegetation and to reseed with climate-appropriate native vegetation.

The U.S. Forest Service and Bureau of Land Management are important resources that should be consulted in developing an emergency response and postfire management plan for wildfires. Along with NOAA's National Weather Service, these federal agencies can provide critical information for establishing effective wildfire mitigation and response procedures. Another resource is the National Fire Protection Association (NFPA), an international fire prevention advocacy group, whose Firewise Communities program can be implemented to complement or serve as an alternative to a CWPP (see www.nfpa.org or www.firewise.org).

ECOSYSTEM MANAGEMENT

Ecosystem impacts vary by region and locality and involve the effects of rising temperatures, increasing sea level, changing precipitation patterns, and increasing storm intensity. The effects of climate change may increasingly have negative impacts on the quality and quantity of natural resources. More severe droughts stretch water supplies to their limits. Stronger storms as well as sea-level rise threaten to damage and inundate property. Rapidly changing climatic conditions can alter ecosystems, and less precipitation and warmer temperatures can contribute to more frequent and severe wildfires. Sea-level rise has the potential to dramatically affect coastal wetlands, the nursery areas and feeding grounds for many commercially significant fish and shellfish species. Drought may have huge impacts on agriculture, timber production, and riverine fisheries.

Local planners and governments can help ensure that natural resources are managed in ways that mitigate climate change and allow communities to adapt to a changing environment.

General Ecosystem Response to Changes in Local Climate

Changes in temperature, moisture regimes, and other primary climate factors may alter the basic environment to which natural communities and their assemblages of plant and animal species have adapted. These changes may be significant enough that habitat tolerances of an individual species may be exceeded and the population of that species will face physiological stresses that make survival difficult. More commonly, though, changing climate conditions alter the competitive dynamics and interactions among different species. For example, white fir trees may have a competitive edge over black oaks on certain extremely cold sites, but warmer low temperatures may change that advantage.

Often, the response of a plant or animal population to changing climate is to shift its range. The key issue then becomes the availability of a comparable habitat and food sources. Where habitat areas are continuous and climate change is gradual, plants and animals may simply migrate northward, although the speeds at which species can move and their capacities to do so vary tremendously. Where suitable habitat is isolated due to natural circumstances or fragmented by man-made features—or when climate change is abrupt—migration may be difficult or impossible.

Ecologists and conservation organizations are recommending the planning and design of land conservation efforts to protect and restore habitat connectivity in order to facilitate the expected migrations of species and ecosystems.

Storm Surge and Coastal Ecosystems

In coastal areas subject to storm surge, barrier islands, dunes, and wetlands can serve as critical allies in the fight against storm surge damage. The Working Group for Post-Hurricane Planning for the Louisiana Coast was formed after hurricanes Katrina and Rita to examine measures that could be taken to reduce damage from future storms. One of the most critical measures that the group identified was preservation and reconstruction of the natural coastal landscape. The group's final report made the following conclusions:

> Barrier islands, shoals, marshes, forested wetlands and other features of the coastal landscape can provide a significant and

PALM BEACH COUNTY URBAN ECOSYSTEM ANALYSIS
Population: 1,265,293

Palm Beach County, Florida, has identified the value that its urban ecosystem has for carbon sequestration and stormwater retention. It has effectively communicated these benefits to the public and has incorporated protection of these qualities into land development regulations and land-use decisions.

Palm Beach County is the largest Florida county by area. Much of the county is unincorporated, and 45 percent of its population lives in these areas, making the county government an important player in land planning decisions. Palm Beach County is also part of the Everglades ecosystem, and its wetlands and natural areas are integral parts of the county's green infrastructure—including tree canopy, open space and grass, and water—that protects water quality, limits soil erosion, improves air quality, and stores atmospheric carbon. The Palm Beach County Department of Environmental Resources Management manages 35 natural areas encompassing more than 30,000 acres. The 221 square-mile Loxahatchee National Wildlife Refuge, part of the Everglades, is also in Palm Beach County.

The county is often battered by tropical storms and hurricanes that destroy both built property and green infrastructure. In 2004, after hurricanes Francis and Jeanne caused significant tree-canopy loss, Palm Beach County received a grant from the Federal Emergency Management Agency to conduct an Urban Ecosystem Analysis (UEA). The UEA evaluated the county's land cover at two scales during two time periods. First, a hurricane assessment used high-resolution data to measure changes in land cover between 2004 and 2006, when several major hurricanes had come through the region. The high resolution made it possible to distinguish between tree-canopy loss due to hurricane damage and that due to development. The second assessment measured land cover changes using moderate-resolution Landsat satellite imagery taken in 1996 and 2006. This analysis showed trends that were due to development and the impacts that the land cover changes have had on air and water quality and on stormwater runoff (Schwab 2009).

(continued on page 144)

(continued from page 143)

The county wanted to use the results to establish a baseline of canopy cover it could refer to for reforestation programs. Another goal was to make the case for connecting land planning decisions to urban ecosystem conservation. By quantifying the economic benefits that natural systems provide in urban areas, planners and elected officials are able to argue more effectively for their preservation and enhancement.

The UEA findings showed that the loss of vegetative land cover led to a decrease in air and water quality. The loss of tree cover from hurricanes exacerbated the trend from development. Environmental benefits were calculated from both data sets to show exactly how much air or water quality was compromised by the loss of trees and vegetative cover. In each year from 1996 to 2006, Palm Beach County stored 880 billion fewer tons of CO_2 and actively sequestered 6,800 fewer pounds of CO_2 as a direct result of vegetative loss. Additionally, 157 million cubic feet of stormwater retention capacity was lost due to tree canopy decline. Based on the retention requirements for a typical two-year peak storm event, the estimated value of this capacity is $316 million. (The construction cost of stormwater retention ponds is estimated at two dollars per cubic foot; Schwab 2009).

In the higher-resolution data set, the loss in tree canopy from hurricane damage equaled a loss of 1.8 million tons of carbon stored in trees and a loss of 14,000 pounds of carbon sequestered annually. Stormwater retention capacity of 146.2 million cubic feet was also lost. The cost to manage this increased stormwater runoff is estimated at $292.4 million (Schwab 2009).

The county can use these findings in campaigns designed to educate the public about the tangible benefits of urban forests. Quantifiable data are especially important in areas subject to natural disasters like tropical storms and hurricanes, as residents are often fearful of replanting trees. Once they learn the benefits that trees and other vegetation provide and are equipped with additional information about the best species and planting locations for hurricane-prone areas, residents tend to be more supportive of community reforestation efforts (Schwab 2009).

The county can also use these results in conjunction with other materials, such as the conservation element in the county comprehensive plan, to better incorporate the value of natural and urban systems in making land-use decisions.

---

potentially sustainable buffer from wind wave action and storm surge generated by tropical storms and hurricanes. Anecdotal data accumulated after Hurricane Andrew suggest that a storm surge reduction along the central Louisiana coast of about three inches per mile of marsh. In a somewhat analogous event, damages from the 2004 Indian Ocean tsunami were reduced by the presence of an extensive, intact mangrove fringe.

Emergent canopies such as provided by forested wetlands can greatly diminish wind penetration, thereby reducing the wind stress available to generate surface waves and storm surge. The sheltering effect of these canopied areas also affects the fetch over which wave development takes place. Shallow water depths attenuate waves via bottom friction and breaking, while vegetation provides additional frictional drag and wave attenuation and also limits static wave setup. . . .

Geologic features such as barrier islands or the land mass and vegetated canopy associated with marshes and wetlands can block or channelize flow. These areas have increased drag that will slow water velocities and may reduce the speed at which storm surge propagates. Together, these effects can significantly restrict the volume of water that is able to reach back-barrier areas and, consequently, that is available to inundate the mainland. (Working Group for Post-Hurricane Planning for the Louisiana Coast 2006, p. 21)

Protection and restoration of critical coastal landscape features can reduce storm surge impacts from hurricanes, tropical storms, and strong nontropical storms. However, as the following discussion on sea-level rise suggests, climate change may complicate preservation and restoration efforts of these important resources.

Sea-level Rise Effects on Coastal Ecosystems

Sea-level rise may affect many coastal ecosystems through inundation and exacerbate the effects of storm surge from tropical and nontropical storm events. Both inundation and storm surge could increase ecosystems' salinity. Hurricane Rita, for example, generated salinity damage to marsh grasses from its storm surge as far as 25 miles inland (Guidroz, Stone, and Dartez 2007).

The Nature Conservancy has been active in assessing the effects of sea-level rise in eastern North Carolina, where it has extensive holdings of coastal forests and wetlands. The low elevation of this area makes it extremely vulnerable to impacts of sea-level rise, as evidenced in Figures 11.2a–¡b, which envision the effect of a two-foot rise in sea level. The scenario is based on a simple "bathtub" model that takes only topography into consideration; if erosion, soil types, subsidence, wave action, and other factors were accounted for, the extent of the impact would be even more dramatic. The Albemarle/Pamlico area of North Carolina could lose more than one million acres of land area due to sea-level rise over the next century.

Even salt-tolerant ecosystems can be affected negatively by sea-level rise. According to one source:

Ecological collapse of tidal wetlands occurs when marsh grasses cannot accrete fast enough to keep abreast of rising sea level in locations where inorganic sediment inputs are low. Eventually plant productivity decreases because excessive submergence effectively drains carbon reserves thereby reducing peat formation and marshes are converted to unvegetated mudflats. Moreover, rise in ambient temperature, in part from global warming, reduces oxygen concentrations in the water column of eroded marsh embayments rendering them poor habitat for most fish species. (Leatherman and Kershaw 2002)

The Nature Conservancy

Figures 11.4a–b. In the event of a two-foot rise in sea level, existing landforms in the Albemarle–Pamlico Sound area of North Carolina (top) are projected to become submerged to the extent shown below.

Coastal area planners need to be aware of the potential impacts of sea-level rise on critical ecosystems, which can affect wildlife habitat and ecosystem functioning. They can also have negative effects on local commercial fishing and sportfishing industries and tourism based on ecological scenery, for example. Creating an inventory of possible impacts and developing responses to them will help coastal communities adapt to the effects of climate change on their local ecosystems.

Drought Effects on Wildlife Habitats

While occasional drought has the potential to have some beneficial effects for wildlife, such as enhanced seed germination in exposed soil at the water's edge, there are a number of negative short- and long-term effects as well. One source notes:

> The severity of the impacts of drought on wildlife populations can vary from area to area. Generally, less water equates to less food and therefore fewer young for all wildlife (fawns, elk calves, and so on). Small, shallow ponds can dry up completely, affecting aquatic wildlife such as ducks, geese, upland birds, and reducing habitat for waterfowl and other wildlife such as amphibians and insects which provide food for other species. In addition, drought effects can be felt for several years afterwards if groundwater is not recharged.
>
> Conflicts between humans and wildlife can occur as dry conditions reduce wildlife forage for deer, elk and black bears, and these animals are pushed closer to humans to find food. Conflicts with farmers and ranchers can occur

when wildlife are attracted to irrigated crops as more succulent and palatable than drought-stressed native vegetation. This can lead to increased damage claims against the state by agricultural producers.

Effects may be felt beyond a drought year due to fewer young elk or deer being born or a reduction in the number of young salmon growing to adulthood. . . .

Some salmon species spawn in channel margins, side channels and smaller tributaries. Spawning would have to occur in mainstem waters if those other areas are unavailable because of low flows. This could make salmon nests, known as redds, and the eggs incubating in them, more susceptible to bed scour during the fall and winter. (Washington Department of Fish and Wildlife n.d.)

As with the effects of sea-level rise, planners in areas likely to be affected by increased drought conditions need to become familiar with the associated impacts on natural systems in order to develop appropriate adaptive responses. Such responses may include public service announcements about the potential for greater human-wildlife conflicts during drought conditions, for example.

Effects of Higher Temperatures on Freshwater Fish and Shellfish

Rising temperatures may affect streams and rivers. Cold water holds more oxygen than warm water does, so increased water temperatures affect the suitability of habitat for species that have high oxygen demands, like trout and salmon. Numerous species depend on normal or near-normal flow and levels for breeding and feeding; increased evaporation due to higher temperatures has the potential to affect the amount of water in streams and rivers. Higher temperatures may also affect snowmelt, resulting in changes to the timing and amount of water in streams and rivers. In some areas, higher-intensity rainfall events may create timing and volume changes that can negatively affect fish and shellfish populations.

Rivers are often affected by "thermal pollution" from the influx of heated water from electric power plants and runoff from parking lots. During times of low flow, thermal pollution can be exacerbated since there is less cool water to moderate the effects from treatment plants and parking lots. These circumstances underscore the need for stream and river buffers; shaded streams are cooler, and buffers provide both rivers and streams with opportunities to both filter and cool discharged water from land uses.

Planners can help address riparian temperature-change impacts through a variety of means, including using green infrastructure and buffers to delay the delivery of parking-lot runoff into streams and other receiving bodies, which creates time for the runoff to cool, and requiring the maintenance or installation of vegetation that shades streams.

ADDITIONAL PLANNING TOOLS FOR NATURAL RESOURCE MANAGEMENT

A variety of planning tools for natural resource management have been mentioned in this chapter, including community wildfire protection plans, water conservation, green infrastructure, and ecosystem-based management. The following section addresses some additional tools planners can use to enhance the management of natural resources.

Use-Value Taxation

In areas where development is rapidly spreading into rural areas, landowners may be unable to pay increasing taxes as the value of their land rises. Landowners may have to sell their land to developers in order to escape this tax burden. The development of these rural lands decreases their ability to sequester carbon. Local governments concerned with maintaining these areas must provide opportunities for landowners to retain their property without bearing the entire cost of higher taxes.

Many states allow agricultural or forestry land to be taxed at a reduced rate compared to other land uses. Land covered by conservation easements is also typically assessed at a level that reflects the reduction in value associated with the easement.

Under the use-value taxation system, land is taxed based on the income potential of the land in agricultural, forestry, or conservation use rather than at its potential market value for more urban development purposes. Such an approach offers an opportunity to encourage the retention of farmland, forests, and natural areas and their current uses.

Conservation Easements

Conservation easements can be used to maintain open space, agricultural lands, and forested areas in order to mitigate climate change and its effects. Such easements can be for a specific term (10, 20, or 50 years, for instance) or in perpetuity. Like use-value assessments, they provide a financial incentive for landowners to keep their land by lowering taxes but have the added benefit of ensuring that once a parcel is placed into easement, its development is permanently restricted. Consequently, conservation easements are an effective way to ensure the protection of rare species, habitats, agriculture and forestry production, and scenic views, while mitigating climate change.

Placing land into a conservation easement does reduce the tax burden on landowners, but the process of establishing an easement can be time-consuming and expensive. Local governments can encourage the establishment of conservation easements by providing technical and financial assistance to landowners throughout the process of surveying, establishing, and monitoring easements. Governments can coordinate these efforts with land-trust groups that work to protect parcels in the area that have ecological, historical, and scenic value.

Other Tools and Techniques

Other ways to promote conservation of natural resources, agricultural, and forest areas include:

Transfer of development rights (TDR). TDR provisions can preserve property owner investment by allowing the development intensity associated with a particular property to be physically separated from that property. In effect, the allowable development that could potentially be put on an agricultural property or environmentally sensitive area is transferred to a less sensitive area for development purposes.

Soils-based zoning. Zoning regulations can be imposed to limit the amount of development allowed on a range of soil types, such as prime agricultural soils. Such an approach may also need to take into account additional physical circumstances associated with the zoned property (proximity to urban development, access, etc.) in addition to soil types so that a wider range of public concerns are addressed.

Cluster development. Concentration of the development allowed on a property creates the opportunity to preserve the remainder of the property for agricultural, forestry, or natural uses. While there are inherent incentives for such "clustering" (e.g., infrastructure expenses are reduced and potential amenities are created), many communities also provide associated density bonuses and other regulatory incentives because there are public benefits associated with maintaining property in open space, including carbon sequestration.

Urban growth or urban service boundaries. These policy- or service-based boundaries demark urban areas from rural areas and support the preservation of agricultural, forestry, and natural areas by making them unavailable or less palatable for development. Such approaches are especially effective in areas where certain urban services are necessary for virtually any kind of development, such as centralized potable water service in areas where individual wells are not feasible due to cost, availability of groundwater, or quality of groundwater.

Compact development pattern. A compact development pattern in general has the potential to increase opportunities for conservation by simply reducing the amount of land needed for urban and suburban development. See Chapter 6 for more information on compact development.

Economic development. Conservation efforts can be greatly enhanced by economic development programs that are designed to take advantage of conserved lands. Examples include heritage tourism, which allows visitors to experience a rural lifestyle, and ecotourism, which promotes the ecological uniqueness of conserved lands.

SUMMARY

The need for management of natural systems becomes more acute in order to mitigate the effects of increased concentrations of GHGs and to create adaptive responses to possible climate change effects like higher temperatures, changes in the rate and amount of precipitation, and sea-level rise. Planners will be challenged to understand the effects of climate change on local natural resources and create or adjust management initiatives accordingly. Particular benefits of achieving natural resource management through maintenance and enhancement of existing ecosystems and through the preservation of open space and forests have been identified, and a range of planning tools to achieve these benefits have been described. It is apparent that comprehensive, systems-based approaches will be needed for planners to be successful in maximizing these benefits for climate change mitigation and adaptation purposes.

TOOLS FOR GUIDING NATURAL RESOURCES DECISIONS

Visioning and Goal Setting	Set goals for agricultural and forestry management, open space preservation, tree canopy and vegetation cover
Plan Making	Identify priority natural resources to be protected in local plans
	Identify particular local and regional threats resulting from climate change and develop responses
Standards, Policies, and Incentives	Create a local tree protection ordinance or incorporate tree protection measures in existing codes and ordinances
	Incentivize open space preservation and conservation of agricultural and forestry resources
Development Work	Encourage new development and redevelopment that maximizes natural resource protection
Public Investment	Promote public investment in green infrastructure, such as the planting of trees or stream restoration, and in preservation of open space

Appendix: Energy Surveys

ENERGY SURVEY 2005

In August 2005, APA and EESI conducted a national survey to assess the state of planners' capacities, knowledge, and educational needs concerning the integration of energy issues and community planning. Three hundred and seventy-seven responses were received from public-sector planners (62.9 percent), private-sector planners (26.8 percent), planning academics (6.1 percent), planners working for nonprofits (2.4 percent), and other planners (1.9 percent). The survey obtained information on what planners feel are the most effective ways to deliver energy training and education.

Summary of Key Findings

Of the total survey population, nearly 65 percent (243) indicated that general energy issues were "very connected" to planning, and 30 percent (113) indicated they were "somewhat connected." The survey also asked how connected specific planning issues were to energy efficiency and renewable energy. Specific planning issues were ranked as being "very connected" to energy efficiency and renewable energy as follows (multiple selections were allowed):

- Sustainability: 83.6%
- Transportation: 83.0%
- Smart growth: 79.0%
- Environmental protection: 76.7%
- Economic development: 59.9%
- Quality of life: 55.2%
- Affordable housing: 49.6%
- Public health: 39.3%

Members of the population who indicated that planning and energy are "very connected" ranked specific planning issues as being "very connected" to energy efficiency and renewable energy as follows (multiple selections were allowed):

- Transportation: 93.8%
- Sustainability: 93.4%
- Smart growth: 89.7%
- Environmental protection: 88.5%
- Economic development: 75.7%
- Quality of life: 67.9%
- Affordable housing: 63.4%
- Public health: 51.0%

Respondents were also asked about their familiarity with specific energy technologies. Most were "very familiar" with passive solar (31.6%) and hydropower (31.3%), which were followed by wind and solar thermal/hot water (both at 21.5%). Those who indicated planning is "very connected" to energy issues produced this same ranking. For all the technologies listed, however, the respondents overwhelmingly indicated they were either "somewhat familiar" or "not familiar" with most of them. The technologies that the respondents were "not familiar" with were led by distributed generation (66.6%), followed by anaerobic digestion (50.4%) and biomass (48.5%).

Nearly all respondents (93.9%) felt that there is a role for planners in helping communities with energy conservation.

ENERGY SURVEY 2007

In August 2007, the APA/EESI partnership launched a second national survey, this time to assess the integration of climate change and energy issues into community planning.

One thousand one hundred and three responses were received from public-sector planners (59 percent), private-sector planners (28.3 percent), planning academics (2.8 percent), planners working for nonprofits (4.1 percent), and other planners (5.8 percent). This split is roughly similar to that in the 2005 survey, but the number of responses increased 293 percent from the 2005 survey.

The 2007 survey included a number of questions that were asked in the 2005 survey, in order to gauge changes since then. It also included new questions regarding climate change policies and plans and general attitudes on the topic.

The following are 10 "big picture" points gleaned from comparing the 2005 to the 2007 survey results.

- *Awareness of energy issues increased among planners.* Of the planners surveyed, 98 percent believed that energy issues are connected to planning, up from 91 percent. Across the board, more planners felt that energy issues were "very connected" to a number of planning issues, including transportation, quality of life, economic development, and public health.

- *Planners were more likely to have primary responsibility in their jurisdictions for developing energy plans and policies.* In 2005, only 19 percent of planners surveyed said their department had jurisdiction over energy concerns; in 2007, 32 percent said the planning department has this responsibility.

- *The number of communities with energy policy statements or plans was poised to grow.* While the percentage reporting existing energy policies or plans stayed the same at about one-quarter, the number of communities with policies in development grew by 10 percent from 2005.

- *Climate change became the top motivator for communities to address energy concerns, and citizen interest continued to be a major factor.* Among planners surveyed, 63 percent cited climate change concerns as a community motivator for developing an energy policy, up from 25 percent in 2005. High utility bills were cited by only 28 percent of planners in 2007, down from 85 percent. Citizen interest remained the second most-cited motivating factor, at 58 percent in 2007.

- *The biggest obstacles to moving forward on energy planning actions were a lack of political interest and the complexity of the issue.* When asked to identify the hurdles facing their locality's energy planning, planners most often cited a lack of political interest (28 percent) and the complexity of energy issues (19 percent). These numbers were practically unchanged from 2005. The importance of citizen interest, however, suggests popular support for energy issues.

- *Reducing demand through energy efficiency was the most common tool that communities planned to use in addressing energy issues.* Of planners whose communities had energy policies, 86 percent indicated that reducing demand through energy efficiency is a key component. Renewable energy purchasing was cited by 65 percent, and reducing peak demand was noted by 43 percent (up 19 percent from 2005).

- *Planners knew more about alternative energy topics and technologies than they did in 2005, but additional education was needed.* More than one quarter of planners surveyed considered themselves "very familiar" with topics such as passive solar, biomass, cogeneration, fuel cells, anaerobic digestion, and distributed generation, but most were not as clear on other technologies such as photovoltaics, wind, transportation biofuels, and geothermal energy. In addition, only a quarter of planners surveyed were familiar with APA's adopted policy guide on energy.

- *A growing number of communities were seeking out energy technology businesses as part of their economic development strategy.* More than one quarter of planners surveyed in 2007 reported economic development plans that encourage energy technology businesses, up from 12 percent in 2005.

- *Most communities had not yet integrated energy concerns into their zoning ordinances and development review procedures.* While 40 percent of planners surveyed said their communities had or were working on energy policy statements, more than 80 percent said their community's zoning ordinance did not address issues such as wind farms, green roofs, solar easements, or distributed-generation energy systems. Few planners reported applying any sort of energy demand standard to site plan review regulations, and only 28 percent said their community had energy efficiency guidelines for housing developments.

- *Most communities did not yet offer incentives for energy efficient or green development, and most planners were not familiar with private market incentives for green building in their area.* Only 7 percent of planners surveyed reported that density bonuses were used to encourage energy elements in new development, while 19 percent reported other incentives to encourage green building. Nearly three-quarters of planners surveyed did not know if other incentives, such as energy- or location-efficient mortgages, were available in their communities. These results are very similar to those of 2005.

References

American Institute of Architects (AIA). n.d. "Architects and Climate Change." Available at www.aia.org/aiaucmp/groups/aia/documents/pdf/aias078740.pdf.

American Solar Energy Society. 2009. *Estimating the Jobs Impacts of Tackling Climate Change*. Prepared by Management Information Services. October. Available at www.ases.org/pdf/ASES_TCCJobs_Summary.pdf.

American Wind Energy Association (AWEA). 2009. *Market Update Fact Sheet*. Available at www.awea.org/pubs/factsheets/Market_Update_Factsheet.pdf.

———. n.d. *How Much Does Wind Energy Cost?* Available at www.awea.org/faq/wwt_costs.html.

Ander, G. A. 2008. "Windows and Glazing." In National Institute of Building Sciences, *The Whole Building Design Guide*. Available at www.wbdg.org/resources/psheating.php.

Appalachian State University. 2007. *Factsheet: Microhyro*. Western North Carolina Renewable Energy Initiative (WNCREI). January. Available at www.wind.appstate.edu/reports/microhydro_factsheet.pdf.

Asheville (North Carolina), City of, and Municode. 2009. *Ordinance No. 3812*. Available at http://library1.municode.com/default-test/home.htm?infobase=12499&doc_action=whatsnew.

Blasing, T. J. 2009. *Recent Greenhouse Gas Concentrations*. Oak Ridge National Laboratory, Carbon Dioxide Information Analysis Center. July. Available at http://cdiac.ornl.gov/pns/current_ghg.html.

Bordoff, J. E., and P. J. Noel. 2008. "Pay-As-You-Drive Auto Insurance: A Simple Way to Reduce Driving-Related Harms and Increase Equity." Hamilton Project discussion paper. July. Washington, D.C.: Brookings Institution. Available at www.brookings.edu/papers/2008/07_payd_bordoffnoel.aspx.

Brown, M. A., et al. 2008. *Shrinking the Carbon Footprint of Metropolitan America*. May. Available at www.brookings.edu/~/media/Files/rc/reports/2008/05_carbon_footprint_sarzynski/carbonfootprint_report.pdf.

Buckley, B., E. J. Hopkins, and R. Whitaker. 2004. *Weather: A Visual Guide.* Tonawanda, N.Y.: Firefly Books.

Bullis, K. 2006. "Cheap, Superefficient Solar." *Technology Review*. November 9. Available at www.technologyreview.com/Energy/17774/?a=f.

California Air Pollution Control Officers Association (CAPCOA). 2009. *Model Policies for Greenhouse Gases in General Plans: A Resource for Local Government to Incorporate General Plan Policies to Reduce Greenhouse Gas Emissions*. June. Available at www.capcoa.org/modelpolicies/CAPCOA%20Model%20Policies%20for%20Greenhouse%20Gases%20in%20General%20Plans%20-%20June%202009.pdf.

California Climate Action Registry. 2009. *General Reporting Protocol*. Available at www.climateregistry.org/tools/protocols/general-reporting-protocol.html.

Cambridge Systematics. 2005. *Traffic Congestion and Reliability: Trends and Advanced Strategies for Congestion Mitigation*. Prepared with the Texas Transportation Institute for the Federal Highway Administration. September 1. Available at http://ops.fhwa.dot.gov/congestion_report.

———. 2009. *Moving Cooler: An Analysis of Transportation Strategies for Reducing Greenhouse Gas Emissions*. Washington, D.C.: ULI.

Chea, T. 2008. "Dwindling Salmon Run Shocks Scientists, Dismays Fishermen." Associated Press. February 3.

Chen, A. 2004. "Cool Colors Project: Improved Materials for Cooler Roofs." *EETD News* 5, no. 4 (fall). Available at http://eetdnews.lbl.gov/nl19/eetd-nl19-1-cool.html.

Ciarlo, C. 2009. Testimony before the House Committee on Science and Technology, Subcommittee on Technology and Innovation. March 31. Available at http://gop.science.house.gov/Media/hearings/ets09/march31/ciarlo.pdf.

Communities Committee et al. 2004. *Preparing a Community Wildfire Protection Plan: A Handbook for Wildland-Urban Interface Communities.* March. Available at www.Doukoupil, T. 2009. "French Reds Are Green." *Newsweek.* August 31. Available at www.newsweek.com/id/212134.

Duerksen, C. 2008. "Saving the World Through Zoning." *Planning.* January: 28–33.

Electric Power Research Institute (EPRI). 2002. *Water and Sustainability.* Vol. 4, *U.S. Electricity Consumption for Water Supply and Treatment—The Next Half Century.* March. Available at http://my.epri.com/portal/server.pt?Abstract_id=000000000001006787.

———. 2007. *Assessment of Waterpower Potential and Development Needs.* March. Available at http://my.epri.com/portal/server.pt?Abstract_id=000000000001014762.

Energy Watch Group. 2007. *Coal: Resources and Future Production.* EWG-Series no. 1/2007. March. Available at www.energywatchgroup.org/fileadmin/global/pdf/EWG_Report_Coal_10-07-2007ms.pdf.

Environmental and Energy Study Institute (EESI) and World Wildlife Fund (WWF). 2009. "Corporate Leadership in Reducing Carbon Emissions." Briefing session. March 27. Available at www.eesi.org/032709_corporate.

Ewing, R., et al. 2008. *Growing Cooler: The Evidence on Urban Development and Climate Change.* Washington, D.C.: ULI.

Ewing, R., and F. Rong. 2008. "The Impact of Urban Form on U.S. Residential Energy Use." *Housing Policy Debate* 19(1). Available at www.mi.vt.edu/data/files/hpd%2019.1/ewing_article.pdf.

Firewise Communities. n.d. *Firewise Construction.* Available at www.firewise.org/newsroom/files/FIREWISE_CONSTRUCTION_TIPS.doc.

Florida, R. L. 2002. *The Rise of the Creative Class: And How It's Transforming Work, Leisure, Community, and Everyday Life.* New York: Basic Books.

Fosdick, J. 2008. "Passive Solar Heating." In National Institute of Building Sciences, *The Whole Building Design Guide.* Available at www.wbdg.org/resources/psheating.php.

Foundation for Water and Energy Education. n.d. *How a Hydroelectric Project Can Affect a River.* Available at www.fwee.org/hpar.html.

Geological Survey of Canada (GSC). n.d. "Permafrost." Available at http://gsc.nrcan.gc.ca/permafrost/climate_e.php.

Gibbard, S. G., K. Caldeira, G. Bala, T. J. Phillips, and M. Wickett. 2005. "Climate Effects of Global Land Cover Change." Lawrence Livermore National Laboratory, Geophysical Research Letters, September 6. Available at https://e-reports-ext.llnl.gov/pdf/324200.pdf.

Gilman, N., D. Randall, and P. Schwartz. 2007. "Impacts of Climate Change: A System Vulnerability Approach to Consider the Potential Impacts to 2050 of a Mid-Upper Greenhouse Gas Emissions Scenario." Global Business Network. January.

Global Carbon Project. 2009. "Super-size Deposits of Frozen Carbon in Arctic Could Worsen Climate Change." *ScienceDaily.* July 6.

Gordon, J. 2007. "Thin-Film Solar Technology Could Be Seriously Clobbering Fossil Fuels in Ten Years." February 20. Available at www.treehugger.com/files/2007/02/thinfilm-solar-clobbering-oil.php.

Guidroz, W. S., G. W. Stone, and D. Dartez. 2007. "Sediment Transport along the Southwestern Louisiana Shoreline: Impact from Hurricane Rita, 2005." Reston, Va.: American Society of Civil Engineers (ASCE). Available at http://wavcis.csi.lsu.edu/pubs/070.pdf.

Hamilton, T. 2009. "Biofuels versus Biomass Electricity." *Technology Review*. May 8. Available at www.technologyreview.com/energy/22628/?a=f.

Hanle, L., et al. n.d. "CO_2 Emissions Profile of the U.S. Cement Industry." Available at www.epa.gov/ttnchie1/conference/ei13/ghg/hanle.pdf.

Hansen, J., et al. 2008. "Target Atmospheric CO_2: Where Should Humanity Aim?" *Open Atmospheric Journal* 2: 217–31. Available at www1.ci.uc.pt/pessoal/asobral/index_ficheiros/L1CLI.pdf.

Hayhoe, K., et al. 2004. "Emissions Pathways, Climate Change, and Impacts on California." *Proceedings of the National Academy of Sciences* 101: 12422–27.

Hayhoe, K., D. Wuebbles, and the Climate Science Team. n.d. *Climate Change and Chicago: Projections and Potential Impacts*. Research summary report. Available at www.chicagoclimateaction.org/filebin/pdf/report/Chicago_Climate_Change_Impacts_Summary_June_2008.pdf.

Henrie, M. 2007. "The New Frontier." *Urban Land*. October: 88–89.

Hill, J., et al. 2006. "Environmental, Economic, and Energetic Costs and Benefits of Biodiesel and Ethanol Biofuels." *Proceedings of the National Academy of Sciences* 103(30): 11206–10. Available at www.pnas.org/content/103/30/11206.abstract.

Hu, P., and T. Reuscher. 2004. *Summary of Travel Trends: 2001 National Household Travel Survey*. Washington, D.C.: U.S. Department of Transportation, Federal Highway Administration. December. Available at http://nhts.ornl.gov/2001/pub/STT.pdf.

Interdepartmental Working Group on Climate Change. 2007. *Adaptation to Climate Change in Agriculture, Forestry and Fisheries: Perspective, Framework and Priorities*. Food and Agriculture Organization of the United Nations, Rome. Available at ftp://ftp.fao.org/docrep/fao/009/j9271e/j9271e.pdf.

Intergovernmental Panel on Climate Change (IPCC). 2007a. *Climate Change 2007: Synthesis Report. Contribution of Working Groups I, II and III to the Fourth Assessment Report of the Intergovernmental Panel on Climate Change*. Ed. Core Writing Team, R. K. Pachauri, and A. Reisinger. Geneva, Switz.: IPCC. Available at www.ipcc.ch/pdf/assessment-report/ar4/syr/ar4_syr.pdf.

———. 2007b. *Summary for Policymakers*. In IPCC 2007a. Available at http://www.ipcc.ch/publications_and_data/ar4/syr/en/spm.html.

———. 2007c. *Climate Change 2007: The Physical Science Basis. Contribution of Working Group 1 to the Fourth Assessment Report of the Intergovernmental Panel on Climate Change*. Ed. S. Solomon, D. Qin, M. Manning, Z. Chen, M. Marquis, K. B. Averyt, M. Tignor, and H. L. Miller. New York: Cambridge University Press. Available at www.ipcc.ch/publications_and_data/publications_ipcc_fourth_assessment_report_wg1_report_the_physical_science_basis.htm.

Johnson. G. 2008. "Salt Water Inundation, Fresh Water Flooding." *Weekly Crop Update* 16, no. 9. University of Delaware Cooperative Extension. May 16. Available at http://agdev.anr.udel.edu/weeklycropupdate/?p=191.

Kalkstein, L. S., J. S. Greene, D. M. Mills, A. D. Perrin, J. P. Samenow, and J.-C. Cohen. 2008. "Analog European Heat Waves for U.S. Cities to Analyze Impacts on Heat-Related Mortality." *Bulletin of the American Meteorological Society* 89(1). January.

King County (Washington). 2007. *2007 Climate Plan*. February. Available at http://your.kingcounty.gov/exec/news/2007/pdf/ClimatePlan.pdf.

———. 2008. *King County Comprehensive Plan 2008*. October. Available at www.kingcounty.gov/property/permits/codes/growth/CompPlan/2008.aspx#chapters.

———. n.d. Water and Land Services. Available at www.kingcounty.gov/environment/waterandland.aspx.

Kitzmiller, F. 2007. "Drought Endangers Crops and Energy Supply." *Independent Mail* (Anderson, S.C.). June 7. Available at www.independentmail.com/news/2007/jun/07/drought-endangers-crops-and-electric-supply.

Lail, M. 2007. "By Saving Fuel, Cities and Towns Save Some 'Green' While Being Green." *Southern City* 57(12). December.

Leatherman, S. P., and P. J. Kershaw. 2002. "Sea Level Rise and Coastal Disasters: Summary of a Forum, October 25, 2001, Washington DC." Washington, D.C.: National Academies Press. Available at www.nap.edu/catalog.php?record_id=10590.

Lovins, A. 2005. *Energy End-Use Efficiency*. Available at www.rmi.org/images/other/Energy/E05-16_EnergyEndUseEff.pdf.

Lowe, E. A. 2001. *Eco-Industrial Handbook for Asian Developing Countries*. Prepared for the Environment Department, Asian Development Bank. Available at www.indigodev.com/Handbook.html.

McIlwain, J. 2008. "The Age of Turbulence: Playing Out Housing's Wild Cards." *Multifamily Trends* (January/February): 16.

McKinsey and Company. 2009. *Unlocking Energy Efficiency in the U.S. Economy*. Available at www.mckinsey.com/clientservice/electricpowernaturalgas/downloads/US_energy_efficiency_full_report.pdf.

McLean, M. L., and K. P. Voytek. 1992. *Understanding Your Economy*. Chicago: APA Planners Press.

National Hurricane Center. n.d. "Slosh Model." Available at www.nhc.noaa.gov/HAW2/english/surge/slosh.shtml.

National Integrated Drought Information System (NIDIS). 2007. *The National Integrated Drought Information System Implementation Plan: A Pathway for National Resilience*. June. Available at www.drought.gov/pdf/NIDIS-IPFinal-June07.pdf.

National Renewable Energy Laboratory (NREL). 2007. *Toward a 20% Wind Electricity Supply in the United States*. Available at www.nrel.gov/docs/fy07osti/41579.pdf.

———. 2008. *Biofuels from Microalgae*. Available at www.nrel.gov/biomass/proj_microalgae.html.

National Research Council (NRC). 2009. *Realistic Prospects for Energy Efficiency in the United States*. Washington, D.C.: National Academies Press.

National Science and Technology Council. 2007. *A Strategy for Federal Science and Technology to Support Water Availability and Quality in the United States*. Report of the Committee on Environment and Natural Resources, Subcommittee on Water Availability and Quality. September. Available at www.ostp.gov/galleries/NSTC/Fed%20ST%20Strategy%20for%20Water%209-07%20FINAL.pdf.

National Trust for Historic Preservation. n.d. "Sustainability by the Numbers." Available at www.preservationnation.org/issues/sustainability/sustainability-numbers.html.

Neal, D. 2008. "A Sustainable Future." *Asheville Citizen-Times*. May 18.

Nelson, A. C. 2006. "Leadership in a New Era." *Journal of the American Planning Association* 72(4): 393–407.

Oak Ridge National Laboratory, Buildings Technology Center. n.d. *Benchmarking Your Building's Energy Performance*. Available at http://eber.ed.ornl.gov/benchmark/intro.htm.

Oregon Department of Energy. n.d. *Biomass Energy: Cost of Production*. Available at http://onlinepubs.trb.org/onlinepubs/nchrp/ciaiii.pdf.

Peterson, T. C., M. McGuirk, T. G. Houston, A. H. Horvitz, and M. F. Wehner. 2008. "Climate Variability and Change with Implications for Transportation." Transportation Research Board paper no. 01104981. Available at http://onlinepubs.trb.org/onlinepubs/sr/sr290Many.pdf.

Petit, J. R., et al. 2001. "Vostok Ice Core Data for 420,000 Years." IGBP PAGES/World Data Center for Paleoclimatology Data Contribution Series #2001-076. Boulder, Colo.: NOAA/NGDC Paleoclimatology Program.

Pew Internet and American Life Project. 2008. *Online Shopping*. February 13. Available at www.pewinternet.org/Reports/2008/Online-Shopping.aspx.

Portland Development Commission. 2006. "Housing Services: Transit Oriented Development (TOD) Property Tax Abatement Guidelines." Available at www.pdc.us/housing_services/programs/financial/transit_oriented_development_guidelines.asp.

Renewable Energy Resource Center. n.d. *Investing in Solar Hot Water*. Available at www.rerc-vt.org/shw_investing.htm.

Ritter, B. 2009. "Gov. Ritter Signs 'Solar Ready Homes' Bill." Press release. Available at www.colorado.gov/cs/Satellite/GovRitter/GOVR/1241443584034. The bill itself is available at www.leg.state.co.us/clics/clics2009a/csl.nsf/fsbillcont3/56DAD78B9D26BD5187257539006E7FF9?Open&file=HB1149_f1.pdf.

Rodale Institute. n.d. "More on Global Warming." Available at www.rodaleinstitute.org/gw/more_on.

Rowe, J. 2006. "Here Comes the Sun." *Greensboro News and Record*. December 12. Available at www.proximityhotel.com/newsrecord121206.htm.

San Francisco County Transportation Authority. 2008. *Mobility, Access and Pricing Study*. Available at www.sfcta.org/content/view/302/148.

Schmer, M. R., et al. 2008. "Net Energy of Cellulosic Ethanol from Switchgrass." *Proceedings of the National Academy of Sciences* 105(2) (January 15): 464–69. Available at www.pnas.org/content/105/2/464.abstract.

Schwab, J., ed. 2009. *Planning the Urban Forest: Ecology, Economy, and Community Development*. PAS Report 555. Chicago: American Planning Association.

Schwab, J., and S. Meck. 2005. *Planning for Wildfires*. PAS Report 529/530. Chicago: American Planning Association.

Scott, M. J., M. Kintner-Meyer, D. B. Elliott, and W. M. Warwick. 2007. "Impacts Assessment of Plug-in Hybrid Vehicles on Electric Utilities and Regional U.S. Power Grids, Part 2: Economic Assessment." Pacific Northwest National Laboratory. November. Available at http://energytech.pnl.gov/publications/pdf/PHEV_Economic_Analysis_Part2_Final.pdf.

Searchinger, T., et al. 2008. "Use of U.S. Croplands for Biofuels Increases Greenhouse Gases Through Emissions from Land-Use Change." *Science* 319(5867): 1238–40. February 29. Available at www.sciencemag.org/cgi/content/abstract/1151861.

Solar Buzz Consultancy. 2009. *Solar Price Index*. Available at www.solarbuzz.com/SolarPrices.htm.

Southwest Florida Water Management District. 2003. "Handwatering & Other Low-Volume Irrigation Applicable Year-Round Water Conservation Measures." Available at www.swfwmd.state.fl.us/conservation/restrictions/factsheet_lowvolume.pdf.

State of Michigan. n.d. "Principles of Ecosystem-Based Management." Department of Natural Resources. Available at www.michigan.gov/dnr/0,1607,7-153-10366_11865-31314--,00.html.

Tans, P. n.d. *Trends in Atmospheric Carbon Dioxide: Mauna Loa*. NOAA Earth System Research Laboratory (ESRL). Available at www.esrl.noaa.gov/gmd/ccgg/trends; data available at ftp://ftp.cmdl.noaa.gov/ccg/co2/trends/co2_annmean_mlo.txt.

Transportation Research Board. 2006. *Commuting in America III*. Available at http://onlinepubs.trb.org/onlinepubs/nchrp/ciaiii.pdf.

United Nations Environment Programme (UNEP). 2008. *UNEP Year Book 2008: An Overview of Our Changing Environment*. February. Available at www.unep.org/Documents.Multilingual/Default.asp?DocumentID=528&ArticleID=5748&l=en.

U.S. Climate Change Science Program (CCSP). 2008. *Impacts of Climate Change and Variability on Transportation Systems and Infrastructure: Gulf Coast Study, Phase I.* Subcommittee on Global Change Research. Ed. M. J. Savonis, V. R. Burkett, and J. R. Potter. Washington, D.C.: U.S. Department of Transportation. Available at www.climatescience.gov/Library/sap/sap4-7/final-report/sap4-7-final-all.pdf.

U.S. Department of Agriculture (USDA). 2002. *The Energy Balance of Corn Ethanol: An Update.* Available at www.transportation.anl.gov/pdfs/AF/265.pdf.

U.S. Department of Energy (DOE). 1996. *Working to Cool Urban Heat Islands.* Available at http://eetd.lbl.gov/HeatIsland/PUBS/WORKING/hibroch.pdf.

———. 2007a. Press release. February 28. Available at www.doe.gov/news/4827.htm.

———. 2007b. *High-Performance Home Technologies: Solar Thermal and Photovoltaic Systems.* Building America Best Practices Series, vol. 6. Prepared by Pacific Northwest National Laboratory and Oak Ridge National Laboratory. June 4. Available at http://apps1.eere.energy.gov/buildings/publications/pdfs/building_america/41085.pdf.

———. 2008. *Ocean Wave Power.* Available at www.energysavers.gov/renewable_energy/ocean/index.cfm/mytopic=50009.

———. n.d. "Terrestrial Sequestration Research." Available at www.fossil.energy.gov/programs/sequestration/terrestrial.

U.S. Department of Transportation Center for Climate Change and Environmental Forecasting. 2008. *The Potential Impacts of Global Sea Level Rise on Transportation Infrastructure.* October. Available at http://climate.dot.gov/impacts-adaptations/pdf/entire.pdf.

U.S. Energy Information Administration (EIA). 2005. *Residential Energy Consumption Survey: Energy Consumption and Expenditure Tables.* Available at www.eia.doe.gov/emeu/recs/recs2005/c&e/summary/pdf/alltables1-15.pdf.

———. 2008. *Electricity Retail Price Fact Sheet.* Available at www.eia.doe.gov/cneaf/electricity/page/fact_sheets/retailprice.html.

———. 2009a. *Annual Energy Outlook.* Report #DOE/EIA-0383(2009). March. Available at www.eia.doe.gov/oiaf/aeo/index.html.

———. 2009b. *International Energy Outlook 2009.* Report #DOE/EIA-0484. Chapter 1. Available at www.eia.doe.gov/oiaf/ieo/world.html.

———. 2009c. State Energy Data System (SEDS), 2007 data. Available at www.eia.doe.gov/emeu/states/_seds.html.

U.S. Environmental Protection Agency (EPA). 2006. *Growing Toward More Efficient Water Use: Linking Development, Infrastructure, and Drinking Water Policies.* January. Available at www.epa.gov/smartgrowth/pdf/growing_water_use_efficiency.pdf.

———. 2008. *Nonpoint Source Pollution: The Nation's Largest Water Quality Problem.* Available at www.epa.gov/OWOW/NPS/facts/point1.htm.

———. 2009a. *Fuel Economy Guide.* Available at www.fueleconomy.gov/feg/feg2009.pdf.

———. 2009b. *Inventory of U.S. Greenhouse Gas Emissions and Sinks.* April 15. Available at www.epa.gov/climatechange/emissions/downloads09/GHG2007entire_report-508.pdf.

———. 2009c. *Nitrous Oxide: Sources and Emissions.* July 20. Available at www.epa.gov/nitrousoxide/sources.html.

———. 2009d. *Transportation and Climate.* Available at www.epa.gov/OMS/climate.

———. n.d. *Natural Gas.* Available at www.epa.gov/rdee/energy-and-you/affect/natural-gas.html.

U.S. Environmental Protection Agency (EPA), Climate Leaders. n.d. "Partner Profile: 3M." Available at www.epa.gov/climateleaders/partners/partners/3m.html.

U.S. Environmental Protection Agency (EPA), Office of Atmospheric Programs. 2006. *Excessive Heat Events Guidebook*. EPA 430-B-06-005. June. Available at http://epa.gov/heatisland/about/pdf/EHEguide_final.pdf.

U.S. Forest Service. n.d. "Protecting Residences from Wildfires." Available at www.fs.fed.us/psw/publications/documents/gtr-050/reccomendations.html.

U.S. General Accounting Office (GAO). 2003. "Freshwater Supply: States' Views of How Federal Agencies Could Help Them Meet the Challenges of Expected Shortages." July. Available at www.gao.gov/new.items/d03514.pdf.

U.S. Geological Survey (USGS). 2008. "Substantial Power Generation from Domestic Geothermal Resources." September 29. Available at www.usgs.gov/newsroom/article.asp?ID=2027.

U.S. Global Change Research Program (USGCRP). 2000/2001. *Climate Change Impacts on the United States: The Potential Consequences of Climate Variability and Change. Final Synthesis Team Reports & Newsletter*. Available at www.usgcrp.gov/usgcrp/Library/nationalassessment/17HE.pdf.

———. 2009. *Global Climate Change Impacts on the United States*. New York: Cambridge University Press. Available at http://downloads.globalchange.gov/usimpacts/pdfs/climate-impacts-report.pdf.

U.S. Green Building Council. 2009. "Green Building Research." Available at www.usgbc.org/DisplayPage.aspx?CMSPageID=1718.

———. n.d. "Green Building Facts." Available at www.usgbc.org/ShowFile.aspx?DocumentID=5961.

Washington [State] Department of Fish and Wildlife. n.d. "Drought Planning." Available at http://wdfw.wa.gov/drought/index.htm.

Wheeler, S. 2008. "State and Municipal Climate Change Plans: The First Generation." *Journal of the American Planning Association* 74(4): 481–96.

Williams, G. 2007. "Resorts Prepare for a Future Without Skis." *New York Times*. December 2.

Wilson, A., and R. Navaro. 2007. "Driving to Green Buildings: The Transportation Energy Intensity of Buildings," *Environmental Building News*. September 1. Available at www.buildinggreen.com/auth/article.cfm?fileName=160901a.xml.

Wood, J. H., et al. 2004. "Long-Term World Oil Supply Scenarios: The Future Is Neither as Bleak or Rosy as Some Assert." U.S. Energy Information Administration. Available at www.eia.doe.gov/pub/oil_gas/petroleum/feature_articles/2004/worldoilsupply/pdf/itwos04.pdf.

Working Group for Post-Hurricane Planning for the Louisiana Coast. 2006. *A New Framework for Planning the Future of Coastal Louisiana after the Hurricanes of 2005*. January 26. Available at www.umces.edu/la-restore/New%20Framework%20Final.pdf.

World at Work. 2009. *Telework Trendlines 2009*. February. Available at www.worldatwork.org/waw/adimLink?id=31115.

RECENT PLANNING ADVISORY SERVICE REPORTS

American Planning Association
Making Great Communities Happen

The American Planning Association provides leadership in the development of vital communities by advocating excellence in community planning, promoting education and citizen empowerment, and providing the tools and support necessary to effect positive change.

512. Smart Growth Audits. Jerry Weitz and Leora Susan Waldner. November 2002. 56pp.

513/514. Regional Approaches to Affordable Housing. Stuart Meck, Rebecca Retzlaff, and James Schwab. February 2003. 271pp.

515. Planning for Street Connectivity: Getting from Here to There. Susan Handy, Robert G. Paterson, and Kent Butler. May 2003. 95pp.

516. Jobs-Housing Balance. Jerry Weitz. November 2003. 41pp.

517. Community Indicators. Rhonda Phillips. December 2003. 46pp.

518/519. Ecological Riverfront Design. Betsy Otto, Kathleen McCormick, and Michael Leccese. March 2004. 177pp.

520. Urban Containment in the United States. Arthur C. Nelson and Casey J. Dawkins. March 2004. 130pp.

521/522. A Planners Dictionary. Edited by Michael Davidson and Fay Dolnick. April 2004. 460pp.

523/524. Crossroads, Hamlet, Village, Town (revised edition). Randall Arendt. April 2004. 142pp.

525. E-Government. Jennifer Evans–Cowley and Maria Manta Conroy. May 2004. 41pp.

526. Codifying New Urbanism. Congress for the New Urbanism. May 2004. 97pp.

527. Street Graphics and the Law. Daniel Mandelker with Andrew Bertucci and William Ewald. August 2004. 133pp.

528. Too Big, Boring, or Ugly: Planning and Design Tools to Combat Monotony, the Too-big House, and Teardowns. Lane Kendig. December 2004. 103pp.

529/530. Planning for Wildfires. James Schwab and Stuart Meck. February 2005. 126pp.

531. Planning for the Unexpected: Land-Use Development and Risk. Laurie Johnson, Laura Dwelley Samant, and Suzanne Frew. February 2005. 59pp.

532. Parking Cash Out. Donald C. Shoup. March 2005. 119pp.

533/534. Landslide Hazards and Planning. James C. Schwab, Paula L. Gori, and Sanjay Jeer, Project Editors. September 2005. 209pp.

535. The Four Supreme Court Land-Use Decisions of 2005: Separating Fact from Fiction. August 2005. 193pp.

536. Placemaking on a Budget: Improving Small Towns, Neighborhoods, and Downtowns Without Spending a Lot of Money. Al Zelinka and Susan Jackson Harden. December 2005. 133pp.

537. Meeting the Big Box Challenge: Planning, Design, and Regulatory Strategies. Jennifer Evans–Crowley. March 2006. 69pp.

538. Project Rating/Recognition Programs for Supporting Smart Growth Forms of Development. Douglas R. Porter and Matthew R. Cuddy. May 2006. 51pp.

539/540. Integrating Planning and Public Health: Tools and Strategies To Create Healthy Places. Marya Morris, General Editor. August 2006. 144pp.

541. An Economic Development Toolbox: Strategies and Methods. Terry Moore, Stuart Meck, and James Ebenhoh. October 2006. 80pp.

542. Planning Issues for On-site and Decentralized Wastewater Treatment. Wayne M. Feiden and Eric S. Winkler. November 2006. 61pp.

543/544. Planning Active Communities. Marya Morris, General Editor. December 2006. 116pp.

545. Planned Unit Developments. Daniel R. Mandelker. March 2007. 140pp.

546/547. The Land Use/Transportation Connection. Terry Moore and Paul Thorsnes, with Bruce Appleyard. June 2007. 440pp.

548. Zoning as a Barrier to Multifamily Housing Development. Garrett Knaap, Stuart Meck, Terry Moore, and Robert Parker. July 2007. 80pp.

549/550. Fair and Healthy Land Use: Environmental Justice and Planning. Craig Anthony Arnold. October 2007. 168pp.

551. From Recreation to Re-creation: New Directions in Parks and Open Space System Planning. Megan Lewis, General Editor. January 2008. 132pp.

552. Great Places in America: Great Streets and Neighborhoods, 2007 Designees. April 2008. 84pp.

553. Planners and the Census: Census 2010, ACS, Factfinder, and Understanding Growth. Christopher Williamson. July 2008. 132pp.

554. A Planners Guide to Community and Regional Food Planning: Transforming Food Environments, Facilitating Healthy Eating. Samina Raja, Branden Born, and Jessica Kozlowski Russell. August 2008. 112pp.

555. Planning the Urban Forest: Ecology, Economy, and Community Development. James C. Schwab, General Editor. January 2009. 160pp.

556. Smart Codes: Model Land-Development Regulations. Marya Morris, General Editor. April 2009. 260pp.

557. Transportation Infrastructure: The Challenges of Rebuilding America. Marlon G. Boarnet, Editor. July 2009. 128pp.

558. Planning for a New Energy and Climate Future. Scott Shuford, Suzanne Rynne, and Jan Mueller. February 2010. 160pp.

For price information, please go to APAPlanningBooks.com or call 312-786-6344.